THE WORLD OF
Classic Tractors

1952 Loyd Dragon.

THE WORLD OF
Classic Tractor

Ian M. Johnston

Kangaroo Press

*Dedicated to
Grant and Marie, Shona and James.*

*May their futures be enriched
by being heedful of the past.*

FRONT COVER
1949 Lanz Bulldog Model N owned by Wally and Betty Johnson, Milvale, N.S.W. (Photo: I.M.J.)

REAR COVER
1912 McDonald EB, restored and owned by noted Victorian collector John Kirkpatrick. (Photo: J. Kirkpatrick)

INSIDE FRONT COVER
A 1936 International McCormick Deering T20 Crawler restored by D. Miller of Carisbrook, Victoria. (Photo: D. Miller)

INSIDE REAR COVER
Photo of the author courtesy of the *Manning River Times*

All rights reserved. No part of this book may be reproduced
or transmitted in any form or by any means, electronic or
mechanical, including photocopying, recording or by any
information storage and retrieval system, nor may it be reproduced
in profit or non profit club or association newsletters or promotional brochures,
without written permission from the publisher or author.

© Ian M. Johnston 1997

First published in 1997 by Kangaroo Press
an imprint of Simon and Schuster Australia
20 Barcoo Street Roseville 2069
Printed in Hong Kong through Colorcraft Ltd

ISBN 0 86417 872 7

Contents

Preface 8
Author's notes 9
Chapter 1	**The CO-OP Classics** 11
Chapter 2	**The Clayton Chain Rail** 15
Chapter 3	**The Nebraska Test** 18
Chapter 4	**The Day I Met Harry Ferguson** 25
Chapter 5	**The Daniels Mogul** 31
Chapter 6	**The Sheppard Diesel** 36
Chapter 7	**In Pursuit of Speed** 43
Chapter 8	**The Legend of Heinrich Lanz** 52
Chapter 9	**Fitch No. 1640** 73
Chapter 10	**Buying a Classic Tractor** 83
Chapter 11	**Kubota – the Late-rising Sun** 89
Chapter 12	**David Brown — the Innovator** 93
Chapter 13	**The Fendt Story** 98
Chapter 14	**The Mail Order Graham Bradley** 103
Chapter 15	**Ludwig Simon** 107
Chapter 16	**The Aultman Taylor 30-60** 117
Chapter 17	**The Eye of the Beholder** 121
Chapter 18	**'Home-Grown' Australians** 131
Chapter 19	**Ferguson – What Really Happened** 144
Chapter 20	**The Galloway Farmobile** 149
Chapter 21	**As British as Winston Churchill** 152
Chapter 22	**Early Tractors in Scotland** 161
Chapter 23	**John Deere – Some Rare History** 168
Chapter 24	**The Motor Cable Ploughing Engines** 175
Chapter 25	**The U.R. Riptoff Tractor Company** 187

Chapter 26 **The Italian Affair** 189

Chapter 27 **The English 'Caterpillar'** 192

Chapter 28 **The Gregarious Garner** 195

Chapter 29 **The Era of the Ivel** 198

Chapter 30 **Lesser Known American Tractors** 205

Chapter 31 **Ferguson Model A in Focus** 228

Chapter 32 **A Salesman on a Bicycle** 233

Chapter 33 **Hanomag of Hanover** 236

Chapter 34 **The Rally Scene** 242

Chapter 35 **Other Interesting European Classics** 247

Chapter 36 **The Christmas Tractor** 257

Metric Conversion Table 261

Index of Illustrations 263

A 1935 photo showing the Deere & Co. dealer at Milan, Illinois, taking delivery of a John Deere Model B. (Photo courtesy of Deere & Co., Moline)

Preface

We of today tend to adopt the somewhat presumptuous philosophy that, because our generation discovered the virtues of the silicon chip, we are more intelligent than those who preceded us. As our legions of satellites create celestial traffic jams in our ungodly heavens, we are usually too preoccupied with our own importance to bother reflecting upon the past.

Yet the most dramatic developments in the history of farm mechanisation had their genesis in the 19th century. These were far-reaching developments destined to reshape the economies of nations for all time; developments that could only have originated in the great intellects of individuals who could not have drawn their prophetic speculations from past experiences, for the past belonged to the era of the ox and the wooden hand plough.

The World of Classic Tractors is a recognition of these protagonists of early farm machinery and their contemporaries who, in the new century, continued with the pursuit of improvement and innovation. It is also an acknowledgement of some of the tractormen whose names never achieved prominence, but whose contribution to the tractor spectrum was significant to those around them. They played an important role in the overall picture.

Essentially then, this is a book for those who have a regard for classic tractors and a desire to further an acquaintance with them.

An interesting comparison. A 1911 steam-powered Aveling Porter 8 n.h.p. traction engine parked alongside a 1911 petrol-powered International Mogul with a single cylinder 20 h.p. engine. The Aveling Porter attracted publicity during the great Queensland floods of 1916 when the town of Clermont was all but swept away and the big machine hauled the few surviving buildings to higher ground. Now restored to its original majesty, the Aveling Porter is proudly on display at the new Clermont site. The amazing story of the Mogul is recounted in Chapter 5. (Photo: Andy Plunkett)

Author's Notes

With the release of two tractor books behind me I imagined I was about to enjoy a well-earned period of blissful freedom from the demands of my publisher.

I relished the prospect of again donning overalls, grasping a handful of spanners and reintroducing myself to the grand old 1938 Marshal M whose restoration I had commenced light years ago. I would be able to take a renewed interest in our farming programme at Wherrol Flat. Indeed, I would have time perhaps to merely contemplate the wonders of nature and watch the trees grow.

The problem was – and is – I am hooked on classic tractors. I cherish my association with these heaps of ironmongery that so reek of character. I seem to continually stumble across tantalisingly intriguing historical revelations that should be shared with other enthusiasts. Also, following a close involvement with tractors spanning more than half a century, I am conscious of being the custodian of tractor tales and experiences that should be recorded for posterity.

When or how a decision was taken to write *The World of Classic Tractors* I am not certain. However, following a gestation period of twelve months, it

1922 Austin restored by Alex and Ron Grosser, on display at the Gunnedah Rural Museum. (Photo: I.M.J.)

has become a reality. During this time I spent some weeks, together with the very patient and supportive Margery Daw, journeying throughout the American Midwest farmlands researching rare and interesting classic tractors. Much of the resulting material is included within these pages so that it may be shared with others.

We were privileged to meet scores of welcoming, easy-going grain farmers who were more than pleased to crank up their colourful classics and wheel them into the sunshine from the dim interiors of their high-roofed barns. It was insisted I should climb aboard and experience for myself the individual characteristics of each tractor. This was sheer bliss – invariably followed by a lengthy sojourn around a scrubbed kitchen table cluttered with tractor literature and plates piled high with home cooking.

Upon returning to Australia I posted off dozens of letters to tractor acquaintances around the world, seeking clarification on historical material and/or technical information. They all responded with enthusiasm. Even the corporate tractor conglomerates to which I directed enquiries were generous with their offers of assistance

Fellow enthusiasts in Australia, knowing of my project, passed the word around. This resulted in letters and photographs arriving regularly in the mail with invitations to use the proffered material in the book for all the world to see. It has, alas, been impossible to incorporate all of this within the available space, but many are featured.

The World of Classic Tractors is more encompassing than my two previous books in the sense that quite a number of the tractors examined never made it to Australia. In this regard I am responding to requests and suggestions from many readers and historians.

Not surprisingly, there are some tractor makes that are not included. These have either been mentioned in my previous works or perhaps will appear in the future. Others, of particular interest, are being reviewed again, but in more detail.

All performance figures are presented in the original measurements of the appropriate era. (It is simply not possible to relate kilowatts to a 4-horse

1951 H.S.C.S. R30-35 Serial No. 24021. Part of the Shaw Collection, Westbury, Tasmania. (Photo: Glen Shaw)

team!) Meticulous care has been taken to edit each of the 5000-odd individual items of data. Murphy's law does apply, however, and errors do occur, sometimes appearing as glaring to the reader. I strive for excellence – maybe I will achieve it one day.

My thanks go to all my old and new friends who helped make this book possible. I also extend my gratitude to the thousands of readers throughout the world who have supported my writing in the past. We are all ambassadors to the worthy task of furthering the interest in classic tractors – those grand old clanking machines that are a part of our heritage.

I.M.J.

This historic agricultural scene was photographed in 1937 on a farm near Ayr in Queensland. The front tractor pulling a 5-disc plough is a Hart Parr 16-30. The second tractor is a Renault HI pulling a cambridge roller, closely followed by a Fordson Model F pulling a tyned cultivator. The centre horse team is attached to a grain harvester, which appears to be out of season when related to the cultivating activities of the tractors. (Photo with special permission from the John Oxley Library, Brisbane)

CHAPTER 1

The CO-OP Classics

One of the joys of an involvement with classic tractors is the discovery of their character – an ingredient singularly absent from their modern computer-designed counterparts. To a seasoned addict, the individual character of an old tractor soon becomes apparent. Once the fuel tap and spark have been set (or the blow-lamp lit), the gearshift checked for neutral, the engine either cranked or swung into life, followed by a jolting drive around the homestead, a tractor reveals its own personal likes and dislikes – in other words, its idiosyncrasies or character. This character is mysterious and intangible to those who have not had the good fortune to 'discover' classic tractors. It is beyond their comprehension. These unfortunates merely smile indulgently at the spectacle of an old tractor being fussed over and patted by a proud owner.

The character of the first CO-OP tractors

A classic tractor that perhaps had more than its share of character was the 1936-37 CO-OP. Ten minutes at the controls was all that was required for an indication that *here is something special*.

The 1936 CO-OP 1, as with the other earlier CO-OP models, was distinctive with its 'bug eye' headlamps and triangular cast front piece. This casting served the dual purpose of counterweight and axle support. In the larger tractors the casting sometimes developed fatigue cracks and was replaced by a fabricated steel plate modification. The photo shows Jack Cochran of Morristown, Indiana at the controls. Jack is an authority on CO-OP tractors and provided much of the technical data that follows. (Photo: I.M.J.)

Around 1934 a printing press manufacturer named Duplex Machinery Co. was approached by the Farmers' Union Central Exchange Co-operative of St. Pauls, Minnesota, to establish a factory capable of producing tractors specifically for the Co-operative. The well-respected tractor designer Dent Parrett (a partner in the Parrett Tractor Co.) was hired to design the tractors. Appropriately, it was decided that the new tractors would be named CO-OP.

The first CO-OP model manufactured by Duplex at their new plant at Battle Creek, Michigan, was released in 1936. It was identified as the CO-OP No. 1. The digit 1 related to the tractor's ability to handle a single furrow (or one bottom) mouldboard plough. The engine used to power the squat little tractor was a 4-cylinder Waukesha identical to that used in the Case RC. In the CO-OP it developed 16 belt h.p. and was coupled to a 4-forward-speed Clark transmission.

In 1937 the CO-OP No. 2 was introduced, having the capability of handling a 2-furrow plough. In common with a number of other independent small U.S. tractor manufacturers of the period, Duplex chose a Chrysler/Dodge side valve automotive engine for the No. 2. It was of 201 cu. inch capacity and produced a leisurely 30 belt h.p. Duplex also used Dodge truck rear transmission and steering box components. The Clark gearbox provided 5 forward speeds.

The third in this original series of CO-OP tractors, the CO-OP No. 3, was added to the range in 1938. As the transmission in the No. 2 had proved entirely trouble free, it was considered adequately robust to be matched to a larger Chrysler/Dodge unit. Accordingly the CO-OP No. 3 (with a 3-furrow capability) was basically a No.2 fitted with a 242 cu. inch 40 h.p. Chrysler/Dodge 6-cylinder side valve engine.

The 1937 CO-OP No. 2 restored by Jack Cochran was powered by a Chrysler 6-cylinder side valve 30 h.p. engine. Note the position of the belt pulley. (Photo: I.M.J.)

Jack Cochran is fortunate in having a son, Mark, who shares his father's enthusiasm for CO-OP tractors. The photo shows Mark at the wheel of his 1937 CO-OP No. 3, the largest of the pre-war range. The engine is a Chrysler 6-cylinder side valve 40 h.p. (Photo: I.M.J.)

Non-conformers

CO-OP tractors did not conform with the trend towards unit construction design. The engines were slung low within a channel chassis and coupled to the transmission by an automotive-type tail shaft. The operator sat encompassed within the tractor rather than upon it. The pan seat was located only inches above the differential. The entire configuration of the steering wheel, gear and throttle control, floor pedals and low seat resulted in a surprisingly comfortable driving position. Even when bowling along in 5th gear at a governed 25 m.p.h. (or an ungoverned 45 plus m.p.h.) the operator felt secure and the tractor handled impeccably. The low vision over the long bonnet was not unlike that experienced when driving a vintage sports car.

The magical character of the CO-OP, therefore, was partly due to this sense of intimacy and the feeling that one became a component of the machine. This heady sensation was heightened by the sound of the powerful throb emitted from the 6-cylinder engine.

The three CO-OP tractor models could be ordered in standard form or as row crop or orchard models. The tricycle row crop versions could be purchased with a cleverly designed front cultivator into which the tractor was driven, thus simplifying its attachment. All tractors were equipped with hand-operated turning brakes.

Whilst there are no accurate figures available, it is believed that only about 300 CO-OP tractors were produced prior to the outbreak of World War II. Most of these were sold to farmers west of the Mississippi.

Around 1938 the Farmers' Union Central Exchange Co-operative was restructured and a separate breakaway group, the National Farm Machinery Co-operative, purchased the Ohio Cultivator Co. of Bellview, Ohio, for the purpose of using their factory as a plant for producing tractors. Another splinter group, the Co-operative Farm Machinery Corporation, established a tractor plant at Arthurdale, West Virginia.

The post-war years

The result of this amalgam of co-operative societies was to introduce a number of small production run tractors all claiming to be CO-OPs. The most interesting of these, and closely following the pattern of the original Duplex Models, was the new CO-OP S3, introduced in 1948 and made at Shelby, Indiana. It still featured a 6-cylinder Chrysler/Dodge engine but was now fitted with a down-draught carburettor. Disc brakes were a sophisticated improvement compared with the expanding drum brakes in pre-war models. An interesting alternative to the straight gearbox transmission was the option of ordering the tractor fitted with a Chrysler fluid drive coupling.

An industrial confrontation resulted in some key members of the Shelby CO-OP factory resigning and joining the nearby opposition firm of Harry A. Lowther Custom Manufacturing Co., the manufacturers of Custom tractors. It is likely that as a direct consequence of this loss of skilled manpower, the new CO-OP Model E3, introduced in 1949, was merely a rebadged Canadian-built Cockshutt 30. This was closely followed by the rebadged Cockshutt 20 and 50 tractors. These Canadian-built tractors were technically advanced for their time, but lacked the character of the CO-OP classics of the 1930s.

The 1948 CO-OP No. 3 was given a more modern appearance than the pre-war models and incorporated a higher seating position, thus affording better operator vision. Whilst Jack Cochran agrees it is an improved machine, he claims it lacks the character of the earlier models. (Photo: I.M.J.)

CHAPTER 2

The Clayton Chain Rail

Clayton & Shuttleworth Ltd. of Lincoln, England, a prominent manufacturer of steam engines and threshing machines, added a crawler tractor to its range in 1916, designated the Chain Rail. At that time, the war in Europe was raging and a large number of Britain's farm workers had joined the military and were fighting in the trenches of France. As a consequence, in order to maintain and hopefully increase food production it was essential that British farmers pursue a course of mechanisation to counter the absence of their labourers. Accordingly, Clayton & Shuttleworth was provided with a government war priority contract from the Food Production Department to design and produce 500 crawler tractors. (The firm was already in receipt of a contract to manufacture 400 Sopwith Camel fighter biplanes for the Royal Flying Corps.)

The first Chain Rail

Crawler track technology had been pioneered in Lincolnshire in the 19th century, therefore there was no lack of expertise in designing the tracks,

The distinctive radiator header tank of a 1917 35 h.p. Clayton Chain Rail. (Photo: I.M.J., courtesy the Lincolnshire Museum of Country Life)

running gear and undercarriage for the proposed tractor. Clayton & Shuttleworth, however, developed their own basic but efficient steering mechanism. This was controlled by a conventional steering wheel, as distinct from the usual levers favoured by other manufacturers. The steering wheel released the cone clutch drive on the chosen side whilst the opposite track remained on full power. This effectively provided gentle steering and directional corrections during ploughing. To enable the tractor to turn abruptly, independent foot-operated brakes were fitted to the outer clutch cast housings. The main engine clutch was also controlled by a foot pedal. The enclosed gearbox provided two forward and one reverse speed. The rear axle was driven by a worm drive mounted above the bevel gear type differential.

A 35 h.p. Clayton Chain Rail pulling a 3-furrow plough. Note the mobile workers' hut against the hedgerow. (From an original sales brochure, courtesy W. Johnson)

A 4-cylinder 384 cu. inch petrol/kero Dorman engine provided the power. Its long stroke characteristics produced 35 h.p. at 1000 r.p.m. Under test the 2.8 ton tractor returned a 4600 lbs pull at 2 m.p.h. Pulling a 3-furrow Moline mouldboard plough to a depth of 5.5 inches in barley stubble, the engine consumed 3.36 gallons per acre when running on kerosene.

The 40 h.p. Chain Rail

Following the success of the 35, Clayton & Shuttleworth upgraded the unit to 40 h.p. around 1922. The additional horsepower was obtained by increasing the r.p.m. of the Dorman engine to 1200. The steering wheel now activated the steering clutches and brakes progressively. The band brakes could also be applied jointly by a foot pedal.

The fuel tank was relocated from a hazardous position over the bonnet to the rear of the engine. Not only was this safer and more practical, it also presented the operator with better forward vision. A new twin upholstered bench seat afforded greater comfort and the foot and hand controls were more conveniently placed.

The Clayton & Shuttleworth Ltd. letterhead makes interesting reading. Note the listing of Major H.D. Marshall as a director. William Marshall Sons Ltd. acquired Clayton & Shuttleworth Ltd. in 1929 during a complex refinancing of both organisations. (Document courtesy W. Johnson)

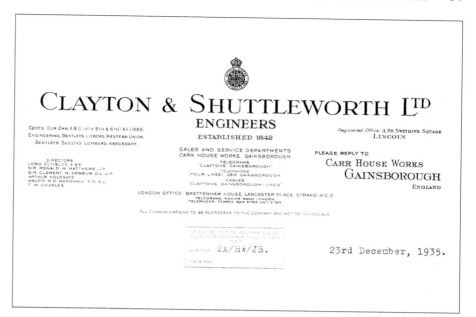

The improved 40 h.p. Clayton Chain Rail. (From an original sales brochure, courtesy W. Johnson)

The profile of the Clayton 40 h.p. Chain Rail. Note the grease cups supplying lubricant to the lower track rollers, front idlers and rear sprockets. The tensioning of the track drive was accomplished by repositioning the twin idler lateral securing bolts and redressing the tension adjusting screw. This moved the idlers forward on the rails, thus tightening the tracks. (From an original brochure, courtesy W. Johnson)

CHAPTER 3

The Nebraska Test

I wish to acknowledge my gratitude to Brent Sampson of the Tractor Testing Laboratory, University of Nebraska, for showing me over the test facility and explaining its functions. My thanks also to Lester F. Larsen (Engineer in Charge for 29 years), Roland Spenst (owner of a remaining Ford B), plus my good friends Vern and Grace Anderson of Lincoln, Nebraska, all of whom provided valuable assistance with my research.

The 'Unethical' Ford

A group of businessmen, led by a financier opportunist named W.B. Ewing, established a tractor manufacturing plant in Minneapolis in 1915. Being aware that 'Ford' was the most identifiable trade name in the U.S.A. and also that it remained unencumbered in relation to farm tractors, Ewing registered his company as 'The Ford Tractor Company'. According to popular hearsay he then employed a labourer named Paul Ford, who was instructed to append his signature to the drawings of the tractor about to enter production. (Author's note: The evidence relating to Paul Ford is based

The only surviving Ford B is owned by Roland Spenst of North Dakota. The photo was taken in July 1995 as it was about to be transported to a local classic tractor show. (Photo courtesy Roland Spenst)

on hearsay but I have been unable to confirm the authenticity of this story.)

The intention was obviously to hoodwink farmers into believing they were being offered a Henry Ford product with the same integrity of design and backup associated with Ford cars. The ethics of such a scam apparently caused no concern for Ewing, who had previously been associated with questionable dealings involving Federal Security Bonds.

Roland Spenst (left) and Vern Anderson check the security of the trailer drawbar. The photo is interesting as it shows the size of the Ford in relation to the men.
(Photo courtesy Vern Anderson)

The Ford Model B (why B?) was released to unsuspecting farmers in mid-1915. It was powered by a marine Gile series M petrol engine developing 16 belt h.p. and with twin horizontally opposed cylinders manufactured by the Gile Boat & Tractor Co. of Ludington, Michigan (later changed to Gile Tractor & Engineering Co. and makers of the Gile Model Q tractor in 1918). The Ford was of 3-wheeled configuration with power delivered to the two front drive wheels through a bull gear and pinion arrangement.

The B was a technical monstrosity. Claims were made by its manufacturer that it was capable of handling two mouldboard ploughs. This could rarely be disputed as the tractor could seldom trundle from the shed to the field without developing a breakdown.

This front view of the Ford Model B shows the marine Gile engine. Note the exposed final drive gear.
(Photo courtesy Roland Spenst)

Wilmot F. Crozier

Unfortunately for Ewing, a Ford B was purchased in 1916 by a Nebraska State Legislator named Wilmot F. Crozier of Polk County. This eminent gentleman was also a prominent grain farmer with considerable influence in government affairs. Crozier's tractor simply would not work. W.B. Ewing seemed disinterested in the problem and could seldom be contacted. After 12 months of being harangued and threatened with a law suit, Ewing reluctantly replaced the delinquent tractor with a brand new one. The second tractor proved no better and, to make matters worse, Ewing disappeared from Minneapolis.

Thoroughly frustrated and angered by the whole episode, Crozier purchased a Big Bull tractor from the Bull Tractor Co., also of Minneapolis, in late 1917. Ironically, his new tractor was powered by the same Gile engine as utilised in the Ford B. This power plant had originally been designed for installation in motor boats. The Bull also had a 3-wheel configuration but with two wheels at the rear, and only the furrow wheel propelled the tractor. (Later models had a modification which provided drive to both rear wheels.)

Congressman W.F.Crozier introduced the Tractor Test Bill in the Nebraska Legislature following his bad experiences with the Ford and Big Bull tractors.

Interestingly, the Big Bull was produced by the makers of Twin City tractors — the Minneapolis Steel & Machinery Co. — on a contractual basis. The original blueprint had been prepared by D.M. Hartsough, the designer of the Emerson Brantingham Big 4. Shortly after Crozier acquired his Big Bull, the Minneapolis Steel & Machinery Co. severed its association with the Bull Tractor Co., as it feared the poor reputation of the Bull could flow back and adversely effect the image of the Twin City tractor.

Crozier's Big Bull proved nearly as unworthy as his Ford B. During his travels around his home State of Nebraska he discovered that many farmers shared his displeasure with unethical and unprincipled tractor producers. It seemed an increasing number of manufacturers were taking advantage of the fact that most farmers had no prior tractor experience. Crozier decided that some stern and immediate action was required.

Crozier recruited the aid of his friend Charles Warner, a Senator from Waverly, Nebraska. Together they sponsored a bill which became law in the State of Nebraska in July 1919. In essence it stated that from that day on, no new tractor could be sold in Nebraska unless it was accompanied by a licence issued by the Nebraska State Railway Commission. In order to be issued with the licence, the tractor manufacturer had to hand the Railway Commission a certificate for each tractor, obtained from the University of Nebraska Agricultural Engineering Department. This could only be acquired after the tractor had undergone a strict testing procedure by the Tractor Test Board established at the University.

The new legislation

The new Nebraska legislation had a dramatic effect on the entire North American tractor industry. Nebraska was a major consumer of farm tractors and few manufacturers cared to contemplate the withdrawal of their marketing operations in that State. The only possible course was to make certain that all their tractor models were submitted to the new Nebraska Testing Laboratory and that they would perform adequately. Thus any farmer considering the purchase of a tractor could first ask to inspect the test result issued by the University of Nebraska Agricultural Engineering Department.

The University of Nebraska Hall of Agricultural Engineering.
(Photo: I.M.J.)

As a flow-on of Crozier's legislation some firms felt they had no option but to terminate their tractor production rather than risk the embarrassment and legal implications of having their claimed performance figures disproved. Others, like Aultman Taylor, were happy to submit their tractors for testing and delighted to find that the relevant performance figures had in fact been conservative.

The Fordson

A 1925 Fordson Model F owned by Mal Brinkman of Hamilton, Vic. The tractor is equipped with a side-mounted Athens disc plough. (Photo: I.M.J.)

An interesting repercussion from Ewing's devious exploitation of the Ford name was that Henry Ford was unable to register his first tractor, released in 1917, as a Ford. This explains why his tractor was named Fordson. It proved to be no handicap as the Fordson F went on to be the all-time top seller of a single model of any farm tractor. Over three quarters of a million units were produced in around 11 years.

The tractor tests

The first tractor tested at Nebraska, on 31 March 1920, was a Waterloo Boy Model N, submitted by Deere & Company, which by that time owned the Waterloo Gas Engine Co. Ten years later, in March 1930, a Twin City 11-20 was examined and given Test No. 175, indicating that in the first 10 years of its operation the facility examined 175 units.

In 1934 notification was sent to tractor firms by the test authority advising that the facility was now equipped to test tractors mounted on pneumatic tyres. Some manufacturers were hesitant to submit tractors thus equipped, as they were not convinced that the future lay in that direction. Allis Chalmers, however, was quick to submit, on 2 May of that year, a Model WC fully mounted on pneumatics. Test No. 223 confirmed a growing belief that in fact pneumatic tyres increased the efficiency of the tractor but that additional ballast was required.

This could readily be achieved by the inclusion of cast iron wheel centres or by adding cast iron counterweights to pressed steel rims. In addition, the tyres could be water ballasted. One of the test engineers is known to have quipped 'Even the farmer's fat wife on the tractor seat would help traction'. Not extremely scientific perhaps, but he made his point. The final tractor tested on steel wheels was a John Deere Model B, Test No. 305, on 6 September 1938. The same tractor was also tested on pneumatics.

The test rig may look old-fashioned, indeed the aluminium front portion dates back over half a century, but it is crammed with computer-controlled high tech wizardry (Photo: I.M.J.)

On 9 April 1940 — Test No. 339 — a Ford 9N fitted with the Ferguson System 3 point linkage and hydraulic draught control was submitted. At that time the test engineers had no method of assessing the benefits of the draught control.

The final tractor submitted prior to the termination of the tractor tests during World War II, was a John Deere AR on 27 October 1941 — Test No. 378. On 22 July 1946 testing resumed when a walk-behind market garden 2-wheeled Bear Cat 3000 was put through its paces.

The drawbar pull of tractors is measured by the compression of this hydraulic cylinder and the results analysed by computer in relation to tractor speed and effort exerted. (Photo: I.M.J.)

Under the direction of the Engineer-in-Charge, Lester F. Larsen, new testing equipment was installed in the post-war years. A new 400 h.p. dynamometer was purchased and tractors were attached to it by their power take off shaft. Belt pulley readings were discontinued in 1979. Additional equipment was designed to measure hydraulic lifting capacities, decibel readings inside cabs and the crash resistance of protective frames. The original gravel test circuit was concreted over when crawler tractors became exempt from the test legislation. The old test laboratory was replaced by a much larger building.

After 29 years of service, Lester Larsen is now retired but has established a tractor museum within the original laboratory building and in an adjacent enclosed courtyard. Tractors are donated to the care of the museum for posterity.

Today testing continues unabated. Manufacturers select the tractors to be tested and must be prepared to state that they are stock models. A factory representative may be present during all stages of the test. The laboratory now has the ability to test engines up to 1000 h.p. and up to 12 000 r.p.m.

This brass plaque is attached to the entrance of the original test laboratory. The building has been classified as a historic site.
(Photo: I.M.J.)

CHAPTER 4

A Personal Recollection
The Day I Met Harry Ferguson

As a young agricultural engineering hopeful attending evening classes at his village in Scotland, the author had the opportunity of meeting the celebrated Harry Ferguson.

Anticipation

Most of the lads in the class were in fact the teenage sons of the district's farmers. Each Wednesday evening, after working arduously all day in the fields, we were bundled off from our respective farms to attend the three-hour agricultural machinery 'tech course'. Our mentor was the local blacksmith and tractor dealer whose name (and I kid you not) was Mr McSpanner. These evenings were a highlight in our young lives as, apart from being informative, they were a welcome respite from the demands of the farm.

On this occasion, we were told to be sure to scrub the cow dung off our boots, put on a clean shirt and be in class at least ten minutes before the normal starting time of 5 p.m. The reason for this unusual attention towards our grooming and punctuality was that a great dignitary had promised to honour us with his presence. Harry Ferguson, the man who had revolutionised the design of the farm tractor, was going to stop off during one of his frequent tours of Scotland and treat us to a specially laid on field day — or more accurately, field evening. In this regard, it should be remembered that daylight remains until well past 10 p.m. on midsummer evenings in these northern latitudes.

Picture the scene. A dozen or so Scottish youths seated at their desks, uncomfortably self-conscious in polished boots and starched shirts, but righteously pleased with their punctuality. That is, with the exception of Tommy McPride, who sneaked in five minutes late owing to his father's bull getting into Miss Sunter's turnips — or so he said!

Mr Ferguson the athlete

Harry Ferguson was not 10 foot tall after all. Rather, he was of medium stature, bespectacled, greying and indeed resembled Mr MacBeth, the village undertaker.

The noticeably overawed Mr McSpanner introduced us to the Great Man. We had previously been warned against raising the fact that he, Mr McSpanner, was the local Fordson tractor agent. So with a threatening reminding glare, Mr McSpanner handed his youthful class over to Mr Ferguson.

Harry Ferguson addressed his somewhat subdued audience with his precise Ulster accent, whilst engaging each of us with a searching eye that demanded that we sit upright and remain alert.

During a 15-minute introductory pep talk about the future of farm mechanisation, he told us how fortunate we were in 1951 to be young, active participants who would play a part in this future as it unfolded. He then informed us that a field had been set aside for the tractor demonstrations at Kingarth Farm and we would adjourn there immediately.

With that he turned on his heel and strode out of the building, leaving us to scramble from behind our desks in hot pursuit. I imagined a Scottish Motor Traction bus would have been waiting outside to transport us the two miles to Kingarth Farm. No such thing. In fact, by this time Harry Ferguson was some distance down the road heading out of the village at the fastest walking pace I had ever seen. The portly Mr McSpanner was doing his best to follow hard on his heels, but to do so was obliged to break into a kind of jog — and this in an era when jogging was considered unseemly and unhealthy! The rest of us hurried along in the rear, fearful of being left behind. This amazing pace was apparently typical of Harry Ferguson, whose mind and body worked only in top gear.

It took around 30 tortured minutes for us cover the distance to Kingarth. Upon arrival, Harry Ferguson looked as fresh and physically under control as when first introduced to us. Mr McSpanner's normal ruddy cheeks were the colour of cooked beetroot and his gasping breath rendered him quite beyond speech. The rest of us were only somewhat less distressed.

Gradually we regained our breath and were able to scrutinise the display which had been prepared for the benefit of our young inquiring minds.

A Ferguson TED on Grangehill Farm, Earlsferry, located in the fertile East Neuk of Fife, Scotland. (Photo: J. Black, 1951)

The cryptic question

The 20-acre field had been adorned with three Ferguson tractors plus a variety of implements. Also two Standard Vanguard vans, looking resplendent in their Ferguson livery, were parked beside a long-wheelbase Albion lorry. The latter had presumably transported the tractors and implements.

We were told by the Great Man that he would give us a verbal run down on the machines and emphasised the importance of taking written notes. There was an immediate mild panic from the group as we realised to our dismay that there was not a pencil or notebook to be found amongst us !

Harry Ferguson was far from pleased. He rendered us a tongue lashing about the necessity to always carry a notebook and pencil as one never knew when an absolutely vital piece of information would come our way which should be written down. (In later years I was to find out that this notebook and pencil requirement was mandatory for all his senior staff.)

Mr McSpanner was as uncomfortable as the rest of us, for he too was unable to produce pen or paper. Eck Reekie was nominated as the obvious choice to sprint back to the village to collect an ample supply of the necessary stationery. He alone boasted legs of such length that he was able to see over the withers of a fully grown Clydesdale and therefore would logically be the best marathon runner of us all. He was dispatched accordingly.

Line drawing taken from a 1951 advertisement showing a TEA Ferguson operating a Ferguson 24-plate disc harrow.

Harry Ferguson asked us what appeared to be a senseless question, quite out of keeping with the sharp mind we were learning to respect. He pointed to one of the tractors to which was fitted a Ferguson 2-furrow mouldboard plough and inquired from us 'At what am I pointing?'

Snotty Pringle, always anxious to impress, quickly retorted 'It's obvious', he smirked, 'a Ferguson tractor'.

'Wrong! Anyone else?' Mr Ferguson snapped inquiringly, his gaze roving over us as he searched for a raised hand.

Well this made us wary, albeit happy with Snotty's obvious bewildered discomfort.

Tooshie Stewart ventured hesitatingly 'A tractor and a plough?'

'You are getting close, boy', in a kinder tone.

I had a reluctant but egotistical feeling that it was up to me, a feeling shared by my peers, as their eyes drifted towards me in expectation. I should explain that I alone in the group had been privileged with a private education in far off Edinburgh and 'appropriately' was afforded a certain respect when it came to worldly affairs. Clearly, this now apparently cryptic question required a college education to supply the likely intellectually complex answer. I was nudged forward.

In actual fact, I had been a singularly dull and inattentive student who had left school, to the relief of my form master, at age 15. Nevertheless I had devoured the print off all known tractor pamphlets since the age of 10 and was able to practically recite their contents from memory— including those relating to Ferguson tractors. The answer to Mr. Ferguson's question was, I hoped, a breeze.

'Mr. Ferguson', I began, outwardly confident but inwardly apprehensive, 'the answer is — you are pointing at the Ferguson System'.

Harry Ferguson gazed benignly upon me whilst Mr McSpanner, quick to note the change in the Ulsterman's demeanour, positively beamed with pride, basking in the aura of having such an 'intelligent' student under his guidance, and no doubt thankful that he hadn't been required to supply the answer.

THE COMPLETE FERGUSON SYSTEM
—*a Ferguson tractor with a Ferguson implement*
gives you these 5 important advantages

1. Penetration without excess built-in weight.
2. Traction without excess built-in weight.
3. Finger-tip and automatic depth control.
4. Tractor's front end stays down.
5. Automatic protection against hidden underground obstructions.

Harry Ferguson Ltd's worldwide advertising in 1951 emphasised the Ferguson System of farming.

The philosophy expounded by Harry Ferguson was that a tractor on its own represents no value. Similarly a plough on its own is valueless. Put the two together and you have a system of farming. The Ferguson System of attaching the implement to the tractor with the Ferguson patented hydraulic design and 3-point linkage resulted in the plough exerting forces which pulled the tractor on to the ground, thereby creating traction that hitherto had only been available with heavyweight, cumbersome tractors.

This breakthrough in tractor implement design was explained and demonstrated to us during the course of the evening.

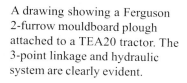

A drawing showing a Ferguson 2-furrow mouldboard plough attached to a TEA20 tractor. The 3-point linkage and hydraulic system are clearly evident.

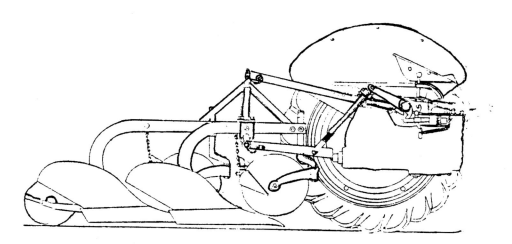

A fine conclusion

Unknown to Harry Ferguson was the fact that, almost to a man, each year the class competed in the East Neuk Annual Junior Ploughing Competition. Consequently, we were no mean hands at ploughing the perfect furrow. Therefore, when volunteers were called to have a go at ploughing with the tractors, Mr Ferguson and his lieutenants were nearly bowled over in the rush.

Never in the history of tractor demonstrations were such excellent straight furrows ploughed. The official Ferguson demonstrator, attired in his gleaming white overalls, who had flaunted his prowess to us just previously, stood and gaped, his mouth open to the elements. He saw his fair attempts at straight furrows being eclipsed by each of the youthful ploughmen. Even young Angus Bull, commonly referred to as 'The Daft Laddie' (every village has one), tore off nonchalantly in second gear, a wide grin upon his countenance, leaving behind an immaculate furrow that had Mr McSpanner's chest all but popping waistcoat buttons.

Harry Ferguson was impressed. He congratulated us on our talent. We three-cheered him and gave an unsolicited solemn undertaking that our fathers would be exhorted to purchase Ferguson tractors. Needless to add, Mr McSpanner (whom it should be remembered was the local Fordson dealer) was out of earshot during such generous commitments. He had instinctively discovered that a surreptitious supply of Tennant's Lager was available at the rear of one of the service vans.

So ended a glorious historic occasion — for me anyway. One that will always remain a nostalgic and memorable experience. And, as I write this chapter, I have come to the final page of my notebook. It will be replaced at once, for never since that distant day in Scotland have I ever been without a notebook and pencil.

CHAPTER 5

The Daniels Mogul

I am deeply indebted to Mr A.J. (Andy) Plunkett for his helpful information which enabled the remarkable story which follows to be told.

The arrival

An important event occurred in the Central Highlands of Queensland in the year 1912. An International Mogul Type C, built the previous year in Milwaukee, Wisconsin, finally arrived at the Daniels property located in the Gindie district south of Emerald. The debut of the big International was of historic significance as it was the very first tractor to be introduced into this predominantly sheep- and cattle-grazing area.

A tractor expert arrived with the Mogul to perform the start-up service. (Note: 'Expert' was the terminology used in Australia, gradually fading out in the 1950s, to describe a technician who specialised in a specific line of machinery, be it tractors, windmills, pumps, harvesters, etc. In the instance of the Gindie Mogul, the expert could not have been all that familiar with his product owing to the uniqueness of *any* tractor in 1912.)

The advent of the new tractor had created a mild sensation in the district. Farmers with their wives and children arrived in horse-drawn buggies from all around to witness the start-up of the new contraption. The general cheerful atmosphere was more reminiscent of the annual Presbyterian Church Picnic.

With much flamboyancy and showmanship the expert proceeded with the ritual of firstly checking the Madison-Kipp force feed lubricator, then tightening bolts and making miniscule adjustments to the friction reverse drive mechanism, and then finally the single-cylinder 20 h.p. Famous engine was primed ready for starting and the choke engaged. With great expectation the silent, fascinated gathering watched as the expert, with exaggerated movements, grasped the flywheel and swung it over. Nothing! With a tolerant smile, such as one might bestow upon a mildly naughty child, the expert again swung the flywheel. Nothing!

After a dozen swings of the flywheel the expert's demeanour changed to that of worried perplexity and by now the sweat poured off him as he leant against the flywheel, recovering his breath. The crowd grew restless and glances were exchanged. No advice was tendered — nobody there had ever before *seen* a tractor, far less understood the mysteries of how to start an internal combustion engine. A couple of older graziers nodded wisely, indicating they had expected something of the sort. Tractors, after all, were nothing more than a passing fancy.

The Mogul proves its worth

Fred Daniels Jnr, the mechanically-minded son in the family partnership, guessing the problem of the obstinate tractor engine, suggested tactfully that it might be worth a try to swing the flywheel in the opposite direction. The expert looked pityingly at Fred, then shrugged his shoulders and, as if to humour him, grasped the flywheel and swung it in the opposite direction. There was a splutter followed by a series of sharp detonations accompanied by clouds of grey acrid smoke until, after a few moments, the big single-cylinder horizontal engine settled down to a rhythmic BANG fiff, fiff, fiff, BANG fiff, fiff,fiff. The crowd cheered and two horses bolted.

No time was wasted in putting the tractor to work. During the early part of the 1912 season the black soil had been baked hard as the result of a prolonged hot, dry spell. Horses could not cope with the harsh conditions but the Mogul was able to haul a converted 6-disc horse plough and rip open the solid ground. When the rains came the ground was fallowed and ready for the seed to be sown, entirely due to the efforts of the new tractor. A bountiful crop was harvested in early October, convincing even the most skeptical that tractors would play an important part in the future of the district's agricultural production.

Above-average rain fell over the Central Highlands in 1913 and, ironically, the weight of the Mogul proved a disadvantage in the wet conditions. The horses were brought in from the back paddock and once again harnessed up. (The excessive weight of early American tractors was a major contributing factor to their lack of acceptance by many farmers. The Mogul Type C was in fact lighter than most but still weighed in excess of six tons. As a consequence, International Harvester, and all the other major manufacturers,

This historic photo shows the Mogul Type C and a binder harvesting the 1912 wheat crop at Gindie. The driver of the Mogul is Fred Daniels Jnr whilst Fred Daniels Snr can be seen operating the binder. Later a second binder was hitched behind the first. Somewhat amazingly, today the remnants of both binders remain at the property. (Reproduced courtesy the Daniels family and the John Oxley Library, Brisbane)

gravitated towards the production of medium and lightweight tractors during the second decade of the 20th century.)

A tragedy

In addition to farming, the Daniels family also owned a contract bore sinking business. The Mogul proved ideal for hauling the Southern Cross boring plant around the adjacent countryside. From 1914 to 1917 it was used almost exclusively for this purpose.

One afternoon in 1917 the Mogul, pulling the boring rig, was en route from Gindie to Lara Park. It was driven by Jim Bertles, a blacksmith and mechanic from Emerald, who worked occasionally for the Daniels. The tractor was rattling along grandly, with Jim enjoying his pipe whilst keeping a watchful eye on the boring rig and portable engine trailing along behind, when suddenly everything went horribly wrong. The steady thump of the Famous engine was instantly transformed into the appalling sound of screaming, tearing metal, which had been preceded by a deafening cannon-like explosion. The big vehicle, with its two trailers, shuddered to a halt. All was silent, apart from the ticking of contracting hot metal.

This Daniels family album photo shows the Mogul pulling the Southern Cross boring plant, with the 6 h.p. Model J 1909 Famous portable engine at the rear. The driver of the tractor is Fred Daniels Jnr., and his passenger is Dan Murphy. (Photo circa 1916)

The investigation which followed revealed that the massive 8.75 inch diameter piston had seized solid in the cylinder. The torque of the huge, whirring flywheels had torn the gudgeon pin from the piston. Disastrously, the flywheels had momentarily kept turning. The result was total mayhem within the engine. It was deemed uneconomical to repair. The tractor was towed ignominiously by a team of horses to the adjacent Yan Yan Station yards, where it was abandoned.

The old hulk was pillaged and raped over the ensuing years by successive generations of machinery repairers who found it an ideal source of nuts and bolts and scrap iron.

The resurrection

What follows is an amazing trail of events. In April 1987 the descendants of the late Fred Daniels, while researching their family history, found the International Mogul featured prominently in the year 1912. References were discovered relating to its demise in the vicinity of Yan Yan Station in 1917.

Accordingly, Larry Daniels and classic engine restorer Andy Plunkett travelled to Yan Yan and tracked down the remains of the old tractor. Lesser mortals would have been daunted by the spectacle of the ancient, rusting machine; however, the pair decided that the Mogul should be retrieved, thus enabling a detailed assessment to be conducted regarding its possible future restoration.

Mr George Sicklinger, the owner of Yan Yan Station, readily agreed to hand over the wreck into the caring hands of the Daniels' descendants. The major items missing included the cylinder (no less), the piston, big end bearing and the roof with its supports. The only one of these items eventually recovered was the cylinder. It was traced to the nearby Caroa Station, where it had been used for years as a forge for heating cattle-branding irons!

This photo of the remains of the Mogul was taken following its discovery precisely 70 years after it had been originally abandoned. (Photo courtesy A. Plunkett)

The restoration of the Mogul commenced at Capella in October 1995. Following six months of an incredibly detailed, difficult and imaginative restoration project, the big tractor was eventually returned to its original glory. The painstaking and arduous work had been executed, under the guiding hand of Andy Plunkett, by numerous members of the Daniels family, plus the generous assistance of scores of other interested parties. It had been a magnificent example of dedicated team work. Today the Mogul Type C looks and sounds as if it came out of the Milwaukee factory yesterday. The restoration was completed in time for the Daniels family reunion in May 1996.

The recovery and restoration of the International Mogul Type C tractor, serial no. TL 2120E and engine no. UB 3620, is an epic saga in the annals of classic tractor dedication. The contribution to the historical record of Queensland's mechanised farming by Andy Plunkett, Fred Daniels' descendants and the other supporters, deserves due recognition and appreciation.

The Emerald Shire Council on the Central Highlands of Queensland, understands the significance of the Daniels Mogul and has draw up plans to erect a display shed for the tractor at Gindie. It will be the centrepiece of a new rural equipment museum.

The 1912 International Mogul Type C fully restored. (Photo: A. Plunkett)

CHAPTER 6

The Sheppard Diesel

In the immediate aftermath of World War II there was a global pressing need for increased food production. Farmers were urged to invest in new tractors and equipment and to convert unproductive land into food-producing acres. The traditional tractor manufacturers were hard pressed to cope with the resultant demand for their products. They were confronted with the massive re-tooling required to switch from military hardware production to that of farm tractors. In Europe bombed out factories had to be completely rebuilt.

The worldwide tractor shortages opened windows of opportunity for newcomers to the arena. Entrepreneurs, often with little or no tractor experience, rushed through engineering drawings and launched a variety of new tractors into the market place. Often these were poorly designed and endured only until the traditional manufacturers were once again able to come on stream with volume production. A few of these opportunistic tractors, however, were in fact well conceived and performed reliably. This is evidenced today by the lineage of those which remain in production.

The Sheppard Diesel was without question one of the more technically interesting tractors that emerged after the war and one which its designers believed could compete favourably alongside such icons as International, Massey Harris and John Deere.

Generators, lifeboat engines and tractors

The R.H. Sheppard Company originated early in the 1930s when Robert Harper Sheppard opened a small manufacturing and engineering business in Hanover, Pennsylvania, specialising in the production of wire fabric used by flour millers for sifting. Noting that there existed an extensive demand by farmers for petrol engine driven generators, the company successfully diversified into that field. Sheppard portable generators were subsequently sold in large numbers into the domestic and export markets.

The principal of the diesel engine became a fascination for Robert Sheppard. By this time his manufacturing business was running smoothly and he was able to allocate time to the development of a number of innovative diesel engine concepts.

The cost of supplying and fitting the Sheppard Diesel engine to the Farmall M was $1200. The example photographed is part of Lynn Klingaman's extensive Sheppard collection. (Photo: I.M.J.)

The end of tractor production

The manufacturing of Sheppard diesel tractors was discontinued in late 1956. During their seven years of production life, sales had been limited not only by the comparatively high price of the tractor, but by the company's rigid policy of extracting full settlement from its dealers for each tractor prior to it leaving the factory. This was at a time when firms such as International Harvester and Massey Harris were offering attractive floor plan schemes enabling their dealers to carry extensive stocks.

Curiously, no Sheppard Diesel was ever tested at the Nebraska Tractor Test Laboratory, a factor upon which competitor tractor manufacturers capitalised.

There are no known Sheppard Diesel tractors in Australia but there is an excellent example of an SD 4 owned by a collector in New Zealand.

Sheppard today

The R.H. Sheppard Company has expanded under the directorship of the founder's son, Mr Peter Sheppard, and has become a major supplier of power steering equipment to the truck and tractor industry. Mr Sheppard is also actively engaged in encouraging the restoration of Sheppard Diesel tractors. In this regard he extends valuable historical research assistance to Mr Lynn Klingaman, who is the President of the Sheppard Diesel Club.

CHAPTER 7

A Personal Recollection
In Pursuit of Speed

The author freely acknowledges that he once harboured a mild obsession for tracking down the world's fastest ever *production* tractor. A psychoanalyst would no doubt have a field day if given the opportunity to probe the origins of this strange cerebral behaviour.

Mr Black's David Brown

It probably all started at Grange Hill Farm when I was a tousle-haired 12-year-old. Mr Black had invested in a sparkling new David Brown Cropmaster. My tractor-driving experience up to that stage was surprisingly extensive, considering I was merely a schoolboy, but this experience had largely been confined to clattering Fordsons (usually on steel cleated wheels) and an assortment of other ageing relics that had been pressed into service during

A 1950 David Brown Cropmaster owned by Col Francis of Collombatti, N.S.W. (Photo: J. Vincent)

the war years. To me the Cropmaster was *awe inspiring*. Even to gaze upon its streamlined magnificence as it sat in the shed conjured up impressions of a speed machine that would not have been out of place at the Lathones Gentlemen's Sporting Car Speed Trials, or perhaps even Brooklands!

There was no doubt that, compared to the others, the Cropmaster was a low flyer. Whilst the Fordson jolted along at a protesting 4 m.p.h., the Cropmaster could whiz between the farms at a breathtaking 12 m.p.h. Adding to its sports car image, it had an upholstered seat that could accommodate two people, albeit with a degree of squeeze. At the impressionable age of twelve I was *certain* this had to be the fastest tractor in all the world.

Mr Gray's Red Farmall

In 1951, at the worldly age of 16 and prior to my departure for the Antipodes, I worked briefly at Brae Head Mains, just outside Edinburgh. This fertile farm was one of several in the Lothians owned by Mr George Gray. Mr Gray did his best to dissuade my father from permitting his youngest son to abscond to the colonies, even though it was to be only for a 2-year 'work experience' duration. He offered instead to take me in hand and give me the necessary training and experience to eventually manage Brae Head or one of his other farms.

I always suspect that the motivation for Mr Gray solemnly putting me in *sole charge* of his new Farmall M was a form of enticement for me to remain in Scotland. He knew very well the allure tractors had for me.

The Farmall M was *some* tractor. Taller, wider, in fact bigger in every way than any other tractor around. It reduced all those ubiquitous fussy grey Fergusons to being like toy tractors. From the lofty elevation of the driving seat I could look down upon all I surveyed, including Mr McNeep who imagined his new Fordson Major E27N was a big machine — probably because previously he had owned a Gunsmith. But the real magic of the Farmall M was experienced when driving it on the main road (totally illegally in my case, as I was still a year off obtaining a driving licence).

A 1952 Farmall M restored by the author and retained as an icon of nostalgia. (Photo: I.M.J.)

According to the operator's manual the M had a top speed of 16.5 m.p.h. However when powering it downhill, sweeping through the village of Cramond Brig, it felt like I was about to go through the sound barrier (which hadn't been discovered yet). Anxious mothers would gather up their bairns as, each afternoon returning from the fields, I careered down the hill with the long bonnet slicing through the wind. There could simply not be a tractor that was faster than this!

My love affair with Charlie

Tucked away in the Riverina district of New South Wales is a sleepy town of great character named Barellan. The tranquillity of Barellan was disturbed during the local Agricultural Show in 1958 when Tail End Charlie was featured in a charity race against a champion trotting horse. For the unenlightened, Tail End Charlie was a Chamberlain Champion tractor that had been specially modified to accompany the cars competing in the rugged Redex — later Mobile — Round Australia Trials. The tractor had a specially tuned Perkins diesel engine plus modified gearing, suspension and brakes. In 1957 it had travelled a whopping 11 140 *miles* over the roughest trails in the world *in nineteen days*. Nearly 600 miles each day!

Tail End Charlie was geared to attain a top speed of 65 m.p.h. The race with the horse was over three laps of the Barellan Show Ground circuit. The agitated horse bolted away from the standing start and left the tractor well behind, but within one lap the bouncing tractor, now in top gear, broadsided past the startled horse in classic speedway style.

Tail End Charlie, Australia's fastest tractor. The photo was taken in 1976 at the premises of the Naracoorte (Vic.) Chamberlain John Deere dealer, Jay Dee Farm Services. On the left is dealer principal John Densley with his son Tim. With them is tractor salesman Barry Johnson. (Photo courtesy John and Jane Densley)

Although at the time of this occurrence I was employed as a field representative by Lanz Australia Pty Ltd, I was on good terms with all my tractor competitors including the chaps from Chamberlain. There was no hesitation, therefore, in granting my request to take Charlie for a spin. The road between Barellan and Narrandera runs due south and for much of the distance is as straight as a die. Despite the windscreen and canopy, the sensation of speed as Charlie and I hurtled south that Saturday afternoon was my most exhilarating tractor experience so far. With my foot planted firmly upon the 'loud' pedal the big tractor roared along the road, much to the open-mouthed amazement of an elderly farmer in his Ford Mainline utility as we rocketed past.

This, I felt, had to be the world's fastest tractor. It probably was — but in a way it was a fraud, because Charlie was far from being a stock standard tractor.

Charlie again

Some years later, in 1964 to be precise, I was to meet up again with Tail End Charlie. As a young sales manager working for Lough Equipment Pty Ltd. of Artarmon, I had lodged a successful tender for the supply of a Hydor A105 air compressor to the Jindalee Shire Council at Cootamundra. It had to be supplied with brackets for mounting to the council's Chamberlain Champion tractor. An approach to my contact at Chamberlain Industries brought a ready response to lend us a Champion for the purpose of measuring up the rear end, thus enabling our workshop foreman to fabricate the brackets in question. On the day it was due to be picked up I was told somewhat apologetically that, as they had no new Champions in stock, the only tractor available for us to borrow was Tail End Charlie, which happened to be parked in their Lidcombe yard.

Normally someone from the workshop would have been dispatched to pick up the Chamberlain, but the prospect of another encounter with Charlie was overwhelmingly persuasive. Having nominated myself for the drive, I took the roundabout way back from Lidcombe, down the Parramatta Road, through the city and over the Sydney Harbour Bridge. What an experience as I manoeuvred the big tractor through the heavy traffic. The spectacle was no doubt surprising — if not alarming — to the conservative North Shore motorists in their Volvos as I powered it up the Pacific Highway through Crows Nest. But they need not have feared, for the Chamberlain handled like the true thoroughbred it was and I was able to temper my enthusiasm with a sense of responsibility.

The French connection

A year later I was introduced to a French nightmare called Latil. This was a 4- wheel-drive, 4-wheel-steer tractor that is frankly better forgotten. A road sealing contractor by the name of Mr Scott had on impulse acquired this — *thing*. He had bought it secondhand at an auction and decided to try it out on the adjacent Epping Highway. Having done so he immediately returned it to his yard and telephoned me, in a state of shock, to arrange its urgent disposal. I was sufficiently naive to readily agree to drive the French tractor to the Lough Equipment stand at the Engineers' Field Day being staged that year at distant Kurnell, in the hope of selling it off the stand.

The stock standard tractor had competed in numerous state fair tractor races which were usually run over a course of 2.5 miles. The black Simpson frequently completed the event 100 yards ahead of its nearest rival. On dirt tracks it regularly *averaged* 60 m.p.h. and the tractor had been clocked at 70 m.p.h. on a bitumen surface (these figures are confirmed in the 'Fremont Guide & Tribune').

I am disappointed in a way that the Friday's status as the fastest production tractor was so short-lived in my mind. For after all, being pulled out of the snow by a tractor of such distinction made a great after-dinner story. On the other hand, I am grateful indeed that the information on the Simpson arrived when it did. I could have received it the day after this book went to press!

I must now go and move my Fordson Major E27N halftrack. It streaks along at 3.5 m.p.h. What bliss!

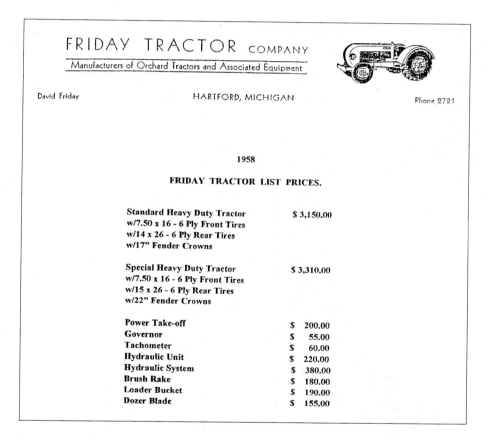

1958 price list of Friday. (Courtesy Rachel Rosenbloom)

CHAPTER 8

The Legend of Heinrich Lanz

The great conglomerates of the stature of the J.I. Case Company of the U.S.A. and William Marshall Sons & Co. of Britain certainly enjoyed enormously high profiles, but there was no organisation that so completely dominated the agricultural machinery realm in its native land than the Heinrich Lanz Company of Mannheim, Germany. Indeed, between 1900 and 1939 the supremacy of the Lanz influence embraced all of Europe including Russia and even extended, in a lesser degree, to the distant lands of Peru, New Zealand, Australia, China, India, Mexico and Syria, to name but a few.

The story of the Lanz dynasty is complex, but made more so by the destruction of records during World War II. I am grateful to my friends in the Lanz Archives, now under the John Deere patronage at Mannheim, and Ulrich Schultz, a colleague whom I first met during the period I spent with Lanz Australia Pty. Ltd. in the 1950s, for sharing their knowledge and recollections with me.

The start of a dynasty

In his youth, Heinrich Lanz had travelled extensively throughout the farmlands of his native Germany and was frequently disturbed by the spectacle of the physically backbreaking labour that was apparently a prerequisite to scratching a meagre living from the soil. He passionately believed that, through his farm machinery business established in 1859, ideas could be implemented which would reduce this physical drudgery and at the same time increase farm output.

Initially the new firm concentrated on carrying out mechanical repairs to the crude farm implements of the period. With expansion in mind, Lanz read of the development work taking place in England by Nathaniel Clayton and Joseph Shuttleworth in the area of portable threshing mills. Sensing that this was something in which his company would like to be involved, he visited the Lincolnshire works of Clayton & Shuttleworth Ltd. in 1860 and arranged to act as German agent for the English firm.

During the visit, Lanz eagerly accepted Nathaniel Clayton's offer to scrutinise the English company's production techniques. He found they had much in common and shared an enthusiasm for the future of steam power in agriculture. Lanz returned to Mannheim in a state of considerable excitement, anxious to introduce new farm mechanisation concepts inspired by his visit to Lincolnshire.

The author and his wife Margery pay their respects to the memory of Heinrich Lanz at the Lanz Memorial Shrine, overlooking the Lanz (John Deere) factory at Mannheim. (Photo: Gunter Gutzlaff)

Three years later, in 1863, the expanding Mannheim firm became involved with the distribution of McCormick reaping machines imported from America. In the meantime Lanz was laying the foundations for the production of a range of farm machinery designed and manufactured in his own engineering shop. Within a few years his plans had become a reality and a steadily increasing range of Lanz machines, including chaffcutters, corn huskers and beet choppers were continually released. The quality of the products could not be questioned, as evidenced by a stream of gold medals awarded at agricultural shows throughout Germany and beyond.

An 1868 award-winning Lanz futterschneidmaschine (chaffcutter). (An original drawing courtesy John Deere-Lanz Archives)

Heinrich Lanz was fully aware that to obtain volume sales for his machinery it would first be necessary to educate farmers to a new way of farming. This in fact presented a daunting challenge. Farmers were accustomed to carrying out their burdensome tasks according to the methods and philosophies handed down for centuries. Accordingly, Lanz embarked upon a costly and unprecedented programme of farm re-education. Special Lanz schools and field demonstrations were established over a period of years throughout farming communities. Farmers and their families were urged to attend on a no-cost basis. As a consequence, Heinrich Lanz and his teams of agricultural experts contributed more to the modernisation of European agriculture than any other single factor. (Note: This was a remarkable achievement never attempted on such a scale by any other farm machinery manufacturer. The nearest approach was when Harry Ferguson established training programs to promote his Ferguson System concept during the early 1950s.)

Lanz also established new standards of industrial awards for his employees. Workers at the Lanz factory received an extra 25% higher hourly rate than the normal rate offered by his competitors. He went on to reduce the standard 10.5 hour working day to 9 hours. Later, a special Lanz hospital was to be constructed to care for his workers and their families.

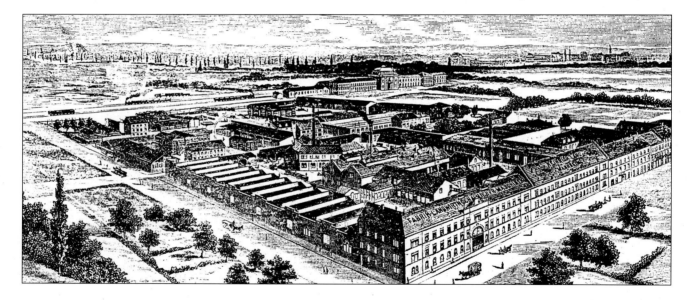

A view of the Lanz factory complex circa 1870. (An original drawing courtesy John Deere-Lanz Archives)

By 1879 the company employed a work force of over 300 men and women. During that year the production of steam engines was commenced. This was a significant milestone in the development of Lanz as the firm went on to become Europe's largest manufacturer of steam engines of all types. There was a huge demand for steam engines in the latter decades of the 19th century. It seemed that everything mechanical was designed to be driven by steam. Farmers, saw millers, engineering shops, flour millers, factories of all types, municipal authorities, military forces — all had an apparently insatiable need for steam engines.

At the Paris exhibition in 1900 Lanz exhibited what was then the world's most powerful steam engine. It was capable of producing 460 h.p., a notable achievement in 1900. In that year the Lanz work force had grown to 2335.

THE LEGEND OF HEINRICH LANZ

An early Lanz steam engine driving a Lanz portable threshing mill. Note the horse shafts attached to the engine. (An original drawing courtesy John Deere-Lanz Archives)

A change in philosophy

Heinrich Lanz turned 64 years of age in 1902. The few other great men in the world who had achieved his status and wealth would likely have been content to ease up the pace and reflect upon the realm which had been created. Lanz, however, intuitively knew that his vibrant and thriving plant, although outwardly and financially successful, had to undergo a change of procedures if it was to maintain its supremacy into the 20th century.

1905 Lanz steam traction engine. Note the support in position for laying back the chimney stack in order to drive under low bridges. (An original drawing courtesy John Deere-Lanz Archives)

For some years Heinrich Lanz had corresponded with the American farm machinery magnate, John Deere. Deere had urged Lanz to undertake the long voyage to the New World and visit his plant in Moline, Illinois, so that the two industrialists could exchange ideas on technology. John Deere died in 1886 and it was not until 1902 that Lanz finally decided to embark on the month-long voyage to Moline, involving coach, rail and a trans-Atlantic crossing. He was welcomed at Deere and Company by Charles Deere, son of the founder. Lanz was fascinated by the production line techniques that he saw being implemented. He noted that the output per worker was almost double that experienced at the Mannheim plant.

Heinrich Lanz returned to Germany and immediately initiated a programme of modernisation involving sweeping changes. Sadly, he was destined not to witness the maturation of his transformations and died on 1 February 1905. The momentum of change within the plant, however, was accelerated when Dr Karl Lanz took control immediately upon his father's death.

New horizons

In 1910 Lanz once again amazed the world when it designed and manufactured a 1000 h.p. steam engine for a Belgian customer. The following year Karl Lanz consolidated his association with the German industrialist Count Zeppelin, who was recognised as the world's leading authority on airships. This resulted in a significant investment by Lanz in a new factory complex specifically structured to build Lanz airships. New jobs were created for around 1800 employees. Apparently the Lanz airships were considered superior to the original Zeppelin units as their frames were constructed from laminated timbers. This was a technique perfected by the Lanz craftsmen and of course the timber frames weighed much less than the iron frames of the Zeppelin airships.

Like his contemporaries around the world, Karl Lanz could foresee that the end of the steam era was drawing nigh. Internal combustion engines were becoming more refined and could produce greater horsepower at a fraction of the weight of steam engines. As a consequence, a team of young designers at Mannheim had been quietly working on a secret project which resulted in the unveiling in 1912 of the first Lanz tractor.

The first Lanz tractors

It is interesting to compare the 1912 Lanz tractor with the units being produced by Marshall of England, International of the U.S.A. and McDonald of Australia. Lanz considered its first Landbaumotor (field working engine) as *part of a system* of farming. To render the *system* complete and functional it was necessary to have the rotary cultivator attached to the rear (this was the principal

1912 Lanz Landbaumotor. The single front wheel and steering device was similar to that used during the same period by the American Wallis Bear tractor.
(An original drawing courtesy John Deere-Lanz Archives)

adopted years later by Howard of Australia with his tractor/cultivator combination and later still, in 1936, by Harry Ferguson with his Ferguson *System*). The Lanz rotary cultivator had originally been designed by a Hungarian and was able to convert a stubble field into a fine tilth seed bed in one pass.

The tractor was powered by a 4-cylinder Kamper petrol engine manufactured for Lanz at the Kamper Berlin factory. It developed 70 h.p. at 600 r.p.m. The gearbox provided forward speeds of 3.6, 5.00 and 6.00 k.p.h. and a reverse of 3.6 k.p.h. Final drive was by an exposed ring gear. The cultivator was powered by two chains driven from a shaft located between the clutch and the gearbox. Somewhat surprisingly, the cultivator was raised and lowered by hydraulic energy.

The Lanz company was in a strong financial situation in 1914. The work force had grown to around 5000 and figures released in 1912 indicated that by that time Lanz had produced no less than 30 000 steam engines, 20 000 mobile threshing mills and 8000 mobile balers. It had also manufactured untold thousands of turnip grinders, wine presses and cream separators in addition to 500 000 chaffcutters and scores of other farm machines.

Owing to the turmoil of World War I and the destruction of Lanz archival material in World War II, there appears to be some doubt regarding the precise date of the introduction of the updated and improved Lanz Landbaumotor. Remaining military photographs show that it was definitely in production during at least part of the 1914-18 War, as it was used as a road haulage tractor for heavy armament and supplies.

It is likely therefore that the new Landbaumotor was released in 1917. It was indeed a modern tractor if one considers the Moguls and Titans still being produced by International Harvester Co., or the Square Turn, Parrett or Rumely. The Jelbart was still frightening horses in Australia and Marshall in England had (temporarily) withdrawn from tractor production entirely.

Now powered by a 4-cylinder 80 h.p. engine of Lanz' own design, the tractor was a true broadacre machine capable of hauling an 8 x 14 inch mouldboard plough to a depth of 14 inches in heavy soil. Its 5.5 ton weight determined that wheel spin was not a problem. The braking system was enclosed within the transmission and activated by a foot pedal.

A 1917 (?) new model 80 h.p. Lanz Landbaumotor fitted with rotary cultivator (An original drawing courtesy John Deere-Lanz Archives)

The Landbaumotor, complete with rotary cultivator, driving a Lanz portable threshing mill. (An original photo courtesy John Deere-Lanz Archives)

A 6-furrow plough being hauled by a Landbaumotor fitted with extra wide wheel grips. (An original photo courtesy John Deere-Lanz Archives)

An interesting photo showing a Landbaumotor doing two jobs in one pass. The seed bed is being prepared with the rotary cultivator and the grain sown directly into the tilth. (An original photo courtesy John Deere-Lanz Archives)

The rotary cultivator in full flight chopping and burying dried potato stocks. (An original photo courtesy John Deere-Lanz Archives)

The first Bulldog

The Landbaumotor was an outstanding tractor and remained in production until 1926. By that time 1500 had been produced at the Mannheim plant. Whilst that represented a reasonable achievement, considering that the price and size of the tractor confined its sales to wealthy landowners, Dr Karl Lanz noted the volume sales being achieved with the Fordson. There was no doubt in his mind that to convince farmers to exchange their draught animals for tractors they had to be offered a reliable and economical low-priced basic machine.

THE LEGEND OF HEINRICH LANZ

Plans were put in place to produce both a single- and 2-cylinder engined new tractor. Priority, however, was directed to the work of Dr Fritz Huber, who was charged with the design of the single-cylinder unit. At the completion of a series of prototype tests, Dr Huber finally unveiled his new tractor in 1921. He had exceeded all expectations. The little tractor was so basic, so simple and so rugged that its squat appearance resembled a Bulldog. Dr Huber in fact referred to it as 'Mein Bulldog'.

The name stuck. The Bulldog was born and destined to become one of the most respected names in the world tractor arena. Sadly, Dr Karl Lanz was never to witness the extent of the great Bulldog phenomenon. He died suddenly at age 48, only weeks after the Bulldog was unveiled.

The brilliance of the Bulldog, designated the HL12, was its simplicity. The engine consisted of 1 cylinder, 1 piston, 1 connecting rod and a crankshaft. That was it! There were no valves, push rods, carburettor or magneto. It could run on virtually any low octane slow combustion fuel, ranging from melted tar to peanut oil, but it was usually more convenient to fuel it with low-priced crude shale oil.

The air fuel mixture in the Bulldog engine was compressed to a ratio of 5.5 to 1 which did not generate sufficient heat for it to detonate. Consequently, built into the cylinder head was a *hot bulb* which had to be heated with a special blowlamp. After around 7 or 8 minutes the hot bulb became red hot. When the fuel mixture was compressed by means of swinging the flywheel

A 1923 scene showing the Lanz Bulldog HL12 driving a Lanz baler by means of a leather endless belt. It is worth mentioning that these early balers were capable of producing magnificent well-packed tight, heavy bales that modern balers can seldom replicate.
(An original photo courtesy John Deere-Lanz Archives)

A page from a 1921 brochure. (Courtesy John Deere-Lanz Archives)

and thus bringing the piston towards the top of its stroke, the red hot bulb caused the mixture to explode. The resultant energy started the engine. There was no need to keep the blowlamp in place as the combustion of the fuel provided sufficient heat for the continual cycling of the piston.

The air for the combustion was drawn (sucked) into the crankcase through a filter, and a simple one-way flap valve prevented it from escaping when the piston returned on its power/exhaust cycle. This action compressed the air and discharged it through a port, now uncovered by the descending piston, into the combustion chamber.

A minute squirt (injection) of fuel was added to the air (now being compressed by the returning piston) and so the process continued. The 2-stroke single-cylinder motor had a 7.48 x 8.66 inch bore and stroke which gave it a capacity of 380.78 cu. inches (6.24 litres). It produced 12 h.p. at a thumping but unruffled 420 r.p.m.

The transmission train was as basic as the engine. A chain from the clutch delivered the engine power to a gear which, in turn, drove the rear wheels by an exposed pinion ring gear drive. In order to reverse the tractor it was necessary to reverse the rotational direction of the engine. This was achieved by cutting off the fuel supply until the engine was on the point of stalling, then throwing the control wide open. With practice this could be accomplished quite expertly.

The cooling system was also basic. A hopper containing water was located on top of the engine. The coolant circulated throughout the water jacketing of the engine by the simple physics of thermal expansion. There were no radiator, no water pump and no belt drives to give trouble. The braking system, however, was possibly the most basic of all. A block of wood was levered against each wheel by depressing a foot pedal. Considering the nuggetty HL12 weighed 2 tons, there presumably were exciting moments when driving the tractor in steep country!

Although only 12 h.p., the Bulldog developed a greater ratio of *torque* than a comparable 12 h.p. petrol or diesel engine. The *semi diesel* Lanz engine had torque (lugging power) characteristics more akin to a steam engine. This was created by the duration of the detonation of the fuel mixture. The burning of the fuel took longer — thus the piston travelling on its power stroke was chased along its journey by the expanding explosive forces. The farmer, probably unaware of *why,* knew that his Bulldog could outpull any other 12 h.p. tractor.

This was a Bulldog idiosyncratic quality that remained with the marque until the production of Bulldogs ceased in 1960. A Bulldog could always outpull any tractor of the same horsepower developed from a conventional petrol or full compression diesel engine.

The Allrad Bulldog and the Felddank

The new Lanz director Dr. Ernst Rochling, who succeeded Dr Karl Lanz, was quick to recognise that the Bulldog had the potential to persuade legions of farmers to change from steam and horses to the advantages of the tractor. The HL12 was inexpensive and became even more so as streamlined manufacturing techniques were developed. Anxious to maintain the Bulldog momentum, Dr Huber and his draughtsmen lost no time and worked around the clock designing successors for the HL12.

In 1923 the brilliance of Dr Huber was again apparent when he introduced the Allrad Bulldog. Identified as the Lanz HP, it had an upgraded but basically similar 12 h.p. engine to the HL12. In every other respect the tractors were different. The Lanz HP featured 4-wheel drive, which in itself was significant, but what made the tractor outstanding was its *articulated chassis*.

A 1923 Allrad Bulldog HP. Note the larger diameter front wheels. (Photo: I.M.J., taken in the staff dining room at the John Deere headquarters, Mannheim)

The tractor was hinged in the centre and steering was accomplished by links and connecting rods which caused the tractor to articulate. (Note: The heavy steering problems which plagued the 4-wheel drive Massey Harris General Purpose seven years later, were avoided by Dr Huber, due to the ingenuity of his articulated design.) The benefits to a farmer of a tractor with 4-wheel drive and articulated steering are obvious, but in 1923 the Bulldog HP was simply years ahead of its time. Conservatively minded farmers were, however, uncomfortable with the new technology and felt more relaxed with the conventional 2-wheel drive proven tractor layout.

Interestingly, the small Italian machinery firm of Pavasi also experimented with a 4-wheel-drive articulated tractor at around the same period, but it failed to eventuate beyond the prototype stage. In the 1950s the concept was rediscovered by a number of European tractor firms, most notably by Holder of Grunbach, Germany, and again in the 1980s by a few of the American companies involved with heavyweight tractors — in particular International Harvester with its 'Snoopy' range of articulates.

For the record it must be stated that the twin-cylinder Lanz was also introduced in 1923 but was discontinued in 1925. Known as the Lanz Felddank, its specially designed Lanz 12.46 litre engine was a vertical 2-stroke valveless hot-bulb semi diesel which developed 38 h.p. at 650 r.p.m. The tractor, weighing around four tons, was provided with 3 forward gears and 1 reverse. Owing to its high price, coupled with the greater appeal of the Bulldog, only 800 Felddanks were produced.

Also for the record, it is interesting to note that whilst the Mannheim factory was busily producing Bulldogs and Felddanks, the Landbaumotor had remained in production. It was not discontinued until 1926, by which time (as previously mentioned) 1500 of the big tractors had been put into operation.

1923 Lanz 2-cylinder Felddank. Its vertical twin-cylinder hot-bulb engine was similar in design to that being used in Swedish Munktells tractors and in the Avance, imported by A.H. McDonald & Co. of Australia. (An original drawing courtesy John Deere-Lanz Archives)

The proliferation of Bulldogs

The rapid expansion of Bulldog models from the time the original HR12 was introduced in 1921 up until the beginning of World War II was phenomenal. It is not the intention to describe each in detail in this publication, as this is a subject for an entire book on its own. (The author commends the

The 1926 HR2 Bulldog was the first of the GROSSER Bulldogs. Producing 28 h.p. from its 10.335 litre engine, it was a massively rugged tractor which still featured the hopper cooling system. It did, however, incorporate 4 forward gears but the engine still had to be reverse cycled to move the tractor rearward. (Photo: I.M.J,. courtesy the Booleroo Steam & Tractor Preservation Society Inc. of South Australia)

64 THE WORLD OF CLASSIC TRACTORS

excellent trilogy of books written by Kurt Hafner, one of which has been translated into English and is available from Kangaroo Press Pty Ltd.) However some of the Bulldog models of this period are highlighted in the following illustrations.

The HR5 15-30 was manufactured between 1929 and 1935. During that time 11 501 were produced. Now with a radiator and fan cooling system plus a reverse gear, the HR5 was a popular tractor in the wheatfields of Australia. (Photo I.M.J., courtesy the Jondaryan Woolshed Museum, Queensland)

The 1930 HR6 Row Crop was yet another example of a European tractor that arrived ahead of its time. The row crop configuration, conceived by International Harvester Co. with the Farmall series, sold strongly in the U.S.A. but European farming was of a different nature and frequently in steep country. As a result, only a few examples of the Row Crop Bulldog were produced. (From an original drawing courtesy John Deere — Lanz Archives)

Le tracteur Lanz à huile lourde
au transport des matériaux de construction.

HEINRICH LANZ MANNHEIM
A.G.

A variety of 20 or 38 h.p. TRAFFIC Bulldogs were released with considerable success during the 1930s. They could be ordered with solid rubber or pneumatic tyres and either single or dual rears. Some were purchased with basic shell mudguards, others were supplied with full length automotive type mudguards. (The above reproduction is from a French sales brochure of around 1932, courtesy John Deere-Lanz Archives)

A 1932 Traffic Bulldog on display in the Lanz Archive Museum at Mannheim. Note the leaf spring under the front axle and the adjacent belt guide. The tractor is equipped with an electric horn and lights, also solid rubber tyres with duals on the rear. (Photo: I.M.J., courtesy John Deere-Lanz Archives)

Many historians give the credit for the first fully enclosed comfort cabin on a farm tractor to Minneapolis Moline with the Model UDLX in 1938. In fact, Lanz produced a superior cabin 4 years earlier in 1934. This, however, was not a Bulldog exclusive as other European manufacturers (e.g. Hanomag) also offered cabins on certain tractors. (An original photo courtesy John Deere-Lanz Archives)

In 1938 an important event took place when Lanz purchased the Hungarian firm of Hofherr Schrantz Clayton Shuttleworth. This company could trace its origins back to 1900 when it commenced the production of thresher machines on a joint venture basis with the English firm of Clayton & Shuttleworth Ltd. H.S.C.S. entered tractor production in the mid-1920s with more than a little assistance from Lanz. The H.S.C.S. tractor was in fact partly designed by Dr Huber and featured a single-cylinder engine similar to that of the Bulldog. The Hungarian factory, however, was unable to match Lanz for production quality control, owing to the outdated machinery used and without the benefit of the exacting Lanz standards.

H.S.C.S. also owned a factory in Austria, which Lanz converted for the production of Lanz threshing machines, grain harvesters and potato planters.

Crawler Bulldogs first appeared in 1936. This original sales brochure advertises the Model D1550 55 h.p. unit equipped with 6 forward and 2 reverse speeds. Weighing 10 250 lbs, it had a drawbar pull of 6832 lbs at 2 m.p.h. (Courtesy John Deere-Lanz Archives)

The war years

Following the outbreak of World War II Lanz continued the production of tractors and farm machinery with a new vigour. Like all nations at war, Germany desperately relied upon farm production. By this time *more than 50% of all tractors working in Germany were Lanz Bulldogs.* (No other tractor firm has ever enjoyed such a national monopoly, with the exception of U.T.B. in post-war communist Romania.)

In 1942 it became law in Germany that all farm tractors had to be converted to run on *holzgas* — wood gas. This involved, among other technical considerations, the fitting of an elaborate and bulky fire box. For tractors fitted with 'conventional' engines such as Fendt, Deutz, Hanomag, etc., this was not a great technical complication, but for the Lanz Bulldog with its valveless hot-bulb scavenging system engine, it was considered impossible. However, some clever engineering overcame the problems and the Holzgas Bulldogs went into production.

The Lanz facilities were pressed into war service. An extensive range of gadgetry had to be designed and manufactured, ranging from V2 rocket components to specially fitted out Bulldog tractors. The tractors were used extensively by the Luftwaffe as aircraft tug tractors and the Wermacht for the movement of heavy armament and munitions.

In the bitterly freezing conditions of the Russian front, charcoal fires had to be kept alight all night under the sumps of petrol truck engines to retain sufficient warmth to enable them to be started the following morning; diesel

The Lanz water tank tower is a familiar landmark in Mannheim. It served as a guiding beacon to allied aircraft during World War II. The photo was taken in the early hours, as the top of the tower was still enveloped in the morning mist.
(Photo: I.M.J.)

The scars from shrapnel and machine gun bullets are still evident in the massive brickwork supporting the water tower.
(Photo: I.M.J.)

engines in tanks had to remain running for weeks on end to prevent freezing. However, the Bulldog tractors could always be started thanks to the normal routine of pre-heating the hot bulb with a blowlamp.

During the last few days of the war the Mannheim plant was almost totally destroyed by allied aircraft. The only major structure left standing was the historic landmark — the Lanz water tower. Today, visitors to the Mannheim plant (now the John Deere works) can easily see the shrapnel and machine gun bullet scars which are clearly evident on the massive brickwork of the tower.

The rebuilding of the Lanz works at Mannheim commenced without delay in 1945, preparing the way for a whole new generation of Bulldogs.

The post-war years

The firm of Heinrich Lanz A.G. never recovered financially from the devastation of World War II. Deere & Co. of Moline, Illinois progressively purchased a controlling interest in the Mannheim plant from 1956 and had totally discontinued all production of Bulldogs by 1962 to make way for the new generation of multi-cylinder John Deere tractors that had been unveiled in the United States in 1960.

The new owners of Mannheim realised they had acquired more than just a tractor plant — they had inherited a legend and a tradition of tractors extending to the far corners of the globe. To the considerable credit of Deere & Co., a Lanz Archive was established within the grounds of the original Mannheim site. It has a staff of caring people who respond to scores of letters received each day from around the world. Adjacent to the climate-controlled, pressurised archive building there is a nostalgic display of early Bulldogs. It is an accepted fact that Lanz Bulldogs are *the most collectable of all classic tractors*. Australia is fortunate in being the largest repository of remaining Bulldogs outside Germany.

The following illustrations offer a review of the diversity of the post-war Lanz tractors.

The 30 h.p. Model L was the smallest of the hot-bulb Bulldogs to be imported into Australia after World War II. This pristine 1949 example is owned by David Miller of Carisbrook, Victoria. (Photo: D. Miller)

The D1706 and its big brothers D2206, D2806, D3206 and D3606 were the first of the new alloy piston-engined Bulldogs introduced in 1952. Fitted with a Bosch electro-magnetic pendulum starter motor, the tractors were started on a petrol/diesel mixture then ran on straight diesel fuel. The hot bulb had become a thing of the past. The compression ratio was raised to 11 to 1, which meant that the new engines were still semi-diesels. Despite retaining all the excellent characteristics of the earlier hot-bulb engines the new units featured a completely redesigned loop scavenging and combustion system. Note the 'auspuff nach unten' — downswept exhaust pipe — which was an option that could be ordered. (Photo courtesy John Deere-Lanz Archives)

The R and T series were introduced in 1955. The most powerful of all Bulldogs, they featured the new alloy piston vibrationless (so-called) engine. The tractors were identical except for engine revolutions and wheel equipment. The R developed 52.1 h.p. at 655 r.p.m. and the T 62.3 h.p. at 800 r.p.m. Deluxe versions were available, known as the DR and DT, which incorporated the addition of a creeper gear range (9 forward and 3 reverse) and independent centralised p.t.o. shaft. Equipped with electric start, the new tractors caused a sensation among Lanz enthusiasts. The tractor pictured is a DT and was photographed at the Lanz stand at the 1957 Sydney Royal Easter Show. (Photo: I.M.J.)

A Lanz Bulldog Model H, with the author driving, opening up virgin country on a farm at Porters Retreat near Oberon, N.S.W. The tractor was transported from there to Orange, where it was awarded the Orange Field Days Award of Merit.
(Photo: L. Simon, circa 1957)

The Model H Bulldog outperformed all comers at the 1957 Orange Field Day, including Ferguson 35, International B250, Fordson Dexta and David Brown 25D. It was the only tractor honoured with the 1957 Award of Merit. The 24 h.p. tractor (along with the 16 h.p. Model C and the 40 h.p. Model Q) had a 12 to 1 compression ratio, the highest of any Bulldog. But it still retained the torque advantages of the semi-diesel principle, enabling it to outperform tractors of equivalent horsepower.
(Photo: I.M.J. — the tractor pictured has been restored by the author and is part of the Chelmsford Collection)

The 40 h.p. Model Q was not available for delivery in Australia until 1959, by which time it was well proven in its native Germany. The Model Q filled the gap between the D2816 (big brother to the Model H, not available in Australia) and the Model R. The majority — but not all — came to Australia painted in the John Deere colours of green and gold. (Photo I.M.J., courtesy the Wheatlands Museum, Warracknabeal)

The John Deere Lanz 440 crawler was the first of the new John Deere tractors to be produced at Mannheim. A few appeared in Australia in late 1959. The 440 was powered by a 3-cylinder Perkins diesel engine which produced 35 h.p. at 2000 r.p.m. The crawler proved troublesome and was replaced in Australia by the American 1010C in 1963. Today the Mannheim plant is one of the most modern in all Europe and produces an extensive range of fine John Deere wheeled tractors which are exported throughout the world. (Line drawing courtesy John Deere-Lanz Archives)

CHAPTER 9

Fitch No. 1640

The Fitch now on public display at the Gunnedah Rural Museum, New South Wales, is one of the most important classic tractor 'discoveries' of recent years. The existence of the original books and correspondence relating to its purchase by Bando Station is also of considerable historic interest. The author expresses his appreciation to Ron Keech, Curator of the Gunnedah Rural Museum, for permission to publish these documents in this chapter.

Mr Bishop's Fitch

When Mr F.E. Bishop of Bando Station, Mullaley, New South Wales, placed an order with the Sydney Auto Truck Co. of Bowen Street, Brisbane, in 1929 for a Fitch tractor, it said something about his character. It would have been much easier to have simply driven into Gunnedah and ordered one of the popular McCormick 15-30 tractors, or perhaps a Ronaldson Tippett Super Drive, or maybe even a Hart Parr Australian Special. Each was readily available, proven in the district and backed by local service and spare parts availability — but Mr Bishop had his mind made up on the Fitch.

Mr Bishop was obviously swayed and impressed by the potential of the Fitch tractor's 4-wheel-drive configuration. Certainly, by 1929 4-wheel-drive tractors had surfaced occasionally in American and Europe during the evolvement of the tractor, but with usually lamentable and brief careers. Apart from the Caldwell Vale in 1910, Australian farmers had little or no experience with such unorthodox machines. Mr Bishop, however, no doubt believed that the Fitch could be the answer to working the sticky self-mulching black soil of the North West Plains without the problem of bogging. This was a tedious common occurrence that plagued conventional steel-wheeled tractors in moist weather.

The Fitch was ordered with the optional centre-mounted grader blade and industrial solid rubber tyred cast wheels, in addition to the standard steel lugged agricultural wheels. Bando Station had an extensive network of tracks to maintain, therefore Mr Bishop foresaw that the grader blade would be of great benefit and receive regular use. Each time cattle or even sheep were mustered and driven by mounted stockmen along the tracks in wet weather, their surface was reduced to a sea of mud. When the hot sun reappeared, the mud baked into treacherous rutted, irregular roadways which only a grader blade could reinstate.

FOUR-DRIVE TRACTOR COMPANY, Inc.

POWER ON ALL FOUR WHEELS
THE ORIGINAL SEVENTEENTH YEAR
Manufacturers of Tractors for Road or Farm
EXPORT DEPARTMENT
15 Moore Street
NEW YORK, N.Y., U.S.A.

Factory
BIG RAPIDS, MICHIGAN

Telephone
WHITEHALL 7622

Cable Address
"RUTOWA," N.Y.

Codes Used
A.B.C. 5th EDITION
LIEBER'S 5 LETTER CODE
BENTLEY'S
also PRIVATE

June 24, 1929.

Mr. F. E. Bishop,
Bando Station,
Mullaley, Gunnedah,
New South Wales, Australia.

Dear Sir:

The factory has referred to us your favor of the 20th of May and note with very much interest that you have Four-Drive Tractor #1640, which we shipped May 1, 1929 to Sydney, Australia. We are glad to send you under separate cover our instruction and parts book for that model tractor.

As regards the parts that you should have on hand, we are submitting a list which would probably be good business to carry in stock. The best suggestions we can give you regarding the servicing of this tractor is to see that it is well lubricated at all times, that there is plenty of water in the radiator, and every two or three week's of steady operation drain out the water, wash out the radiator and the circulation system freely and refuill with fresh water.

Oil level should be kept to a good point which is indicated by the oil indicator on the rear of the motor block, and should be drained at least once a week and refilled with fresh oil.

Once in every two or three months' running, the inspection door should be taken off on the bottom section of the front housing, a bar put under the bottom of the vertical drive shaft and see if there is any considerable play up and down on the vertical shaft. It should have three or four thousandths play but no more.

If you find there is too much play, the cap on top should be removed and take out one or two of the thin paper shims but not too many. Avoid excessive play in your front end; if you don't you will brake up the front end. Just plenty of oil, plenty of cooling and reasonable attention to this tractor will give you perfect satisfaction.

Very truly yours,
FOUR DRIVE TRACTOR CO. INC.

Manager

ML:AC.

The letter addressed to Mr Bishop from the Manager of the New York office of the Four Drive Tractor Company Inc. makes interesting reading. The poor quality of the reproduction is due to the ravages of age upon the original.

Unusual specifications

The Fitch Four Drive Model D4 was indeed an amazing tractor. It was manufactured in Big Rapids, Michigan by the Four Drive Tractor Company Inc. The firm first released its visionary tractors in 1916. The Model D4 was not introduced until 1920, when it became apparent there was a need for a more robust and powerful unit. The new model was powered by a Climax K Series 4-cylinder engine made by the Climax Engineering Co. of Clinton, Iowa. (This was the same power plant as used in the Square Turn — see Chapter 30.) The engine had its 4 cylinders of 5 x 6.5 inch bore and stroke cast in pairs.

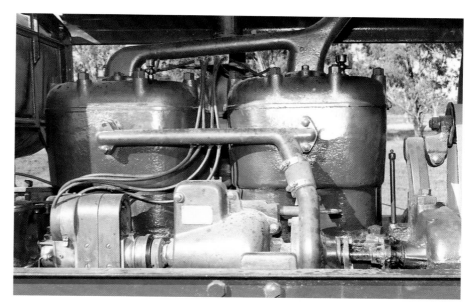

The Climax K Series 4-cylinder engine No. 8422.

The tractor had a dry weight of 3.5 tons when equipped with the cast wheels and solid rubber tyres (36 x 7 inch front and 40 x 7 inch rear) and including the grader blade. Although not confirmed by a Nebraska Test, the engine in the Fitch was claimed by the manufacturers to develop 20 drawbar and 35 belt h.p. at 800 r.p.m., which was delivered to the gearbox via a Borg & Beck clutch. The 3 forward gears were rated at 1.5, 2.5 and 4 m.p.h. and reverse at 1.5 m.p.h. The power to the rear axle was delivered by a Timkin worm drive (similar to the Fordson Model F). The front axle was driven by a clever patented design using a bevel gear principal, so arranged to eliminate any power loss whilst steering into a curve. (It is worth noting that the *entire axle* turned as distinct from the fixed axle of modern 4-wheel-drives with conventional steering and tie rods, necessitating universal or bevel drive at each wheel.)

The steering wheel was connected by the shaft to a worm gear in the steering box. The lateral shaft and pulley extending from the steering box acted as a windlass for the chain connected to the front axle. Unlike most chain windlass steering systems, the Fitch used an intermediate pulley between the steering box and the axle, which served to render the steering considerably more positive.

76 THE WORLD OF CLASSIC TRACTORS

This illustration of the 1929 Fitch at the Gunnedah Rural Museum shows the front casting containing the bevel drive. The canister below the fuel tank is a water air cleaner — it is located in an unfortunate position as it would attract all the dust from the wheels and the grader blade. The lever alongside Roy Morris, the operator, is the left hand control for raising and lowering the left side of the grader blade. (Photo: I.M.J., courtesy Gunnedah Rural Museum)

This right hand view of the Fitch shows the mounting of the grader blade and the location of the belt pulley. (Photo: I.M.J., courtesy Gunnedah Rural Museum)

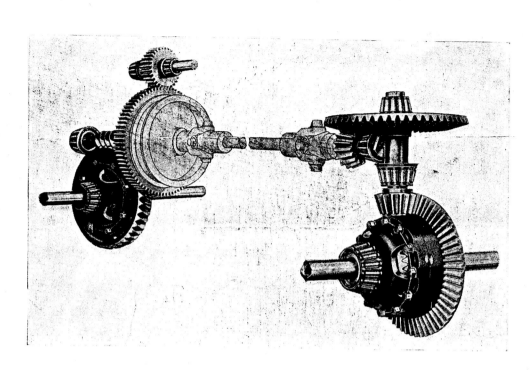

Gears

Note carefully the above photograph. Count the gears. This is all in the FOUR DRIVE TRACTOR. Note the simplicity. Then go over the Crawler type, a two drive. Count the gears and note the excessive loss and wear in friction. Note the complication.

Then go over the two-wheel rear drive—count the gears—note the dead weight pushed ahead on front wheels, which is useless, a dead load, expensive and wasteful, performing no function, but it must be so to compensate in the type of drive.

Indirect complication, squared by direct simplicity, means a saving of 40 per cent. Therefore, save the difference, which is WASTE.

All tractors, regardless of weight or type, on a hard, level surface, each having the same drawbar pull, will carry the same load between terminals. The question is: the cost per-ton-mile by the use of each of the types. If it should cost 10 cents per-ton-mile by the use of the FOUR DRIVE—the Dray Horse with equal strength and power in each of the four legs, and 20 cents per ton-mile by the use of the Crawler type, and 19 cents per ton-mile by the use of the two rear drive types—this difference is profit, in the cents which make dollars and known only as waste.

Those who use the latter two types point to wisdom as a justification. This is one's privilege of course, but, BUT, *BUT*—in the results obtained in the use of each of the three types, measured by the cents which grow into dollars—the difference in cost per ton-mile, the controlling factor, it is then wisdom will assert itself, exercising the right to save the difference, which is common waste, pure and simple—nothing else.

Buy a Four-Drive and Save the Waste

The transmission for the Fitch D4 was designed by the Cotti Transmission Co. of Rockford, Illinois. This descriptive page, reproduced from an original sales brochure, reflects the style of advertising of the period and makes an interesting comparison with the more subtle nature of modern sales promotion.

This interesting photograph was contained in a promotional leaflet put out in 1929 by Sydney Auto Truck Co., the Australian agents for the Four Drive Tractor Company.

FITCH NO. 1640   79

WHY A FOUR DRIVE TRACTOR?

Look again at the front Cover. You would scarcely send an outfit into the field with a fine pair of draft horses and only a pair of saw-horses ahead of them.

No more would you select a Kangaroo hopping on its hind legs only, as a draft animal instead of a mule, which pulls on all four feet.

Nor would you shoe a horse only behind for pulling a load.

The drive on all **FOUR** of the wheels adds another pair of **PULLERS** instead of something to propel as in the case of the saw-horses or idle front wheels of the two wheel drive tractor.

The ground packing tendency is reduced by the fact that the weight is evenly distributed on the four wheels, instead of only on two, as in the case of the two wheel drive or the Kangaroo.

A tractor can deliver at the draw-bar just as much power as it can get traction, less such amount as is required for propelling such portion of its weight as is on idle front wheels. The same as a horse, shod on all four feet, can pull more than one shod behind only.

The Fitch Four Drive Tractor. Maximum Weight (100%) on Traction.

FITCH FOUR-DRIVE TRACTOR, MODEL D-4. 20 H. P. DRAW-BAR. 35 H. P. BELT
(THE ORIGINAL: THIRTEENTH YEAR)

The text contained in this reproduced leaflet is quaint but definitely persuasive.

80 THE WORLD OF CLASSIC TRACTORS

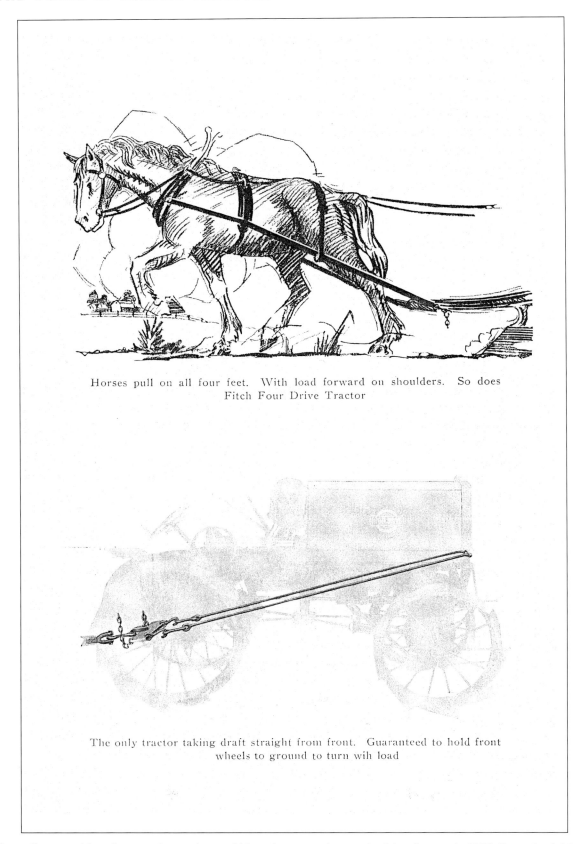

Horses pull on all four feet. With load forward on shoulders. So does Fitch Four Drive Tractor

The only tractor taking draft straight from front. Guaranteed to hold front wheels to ground to turn wih load

The above diagram, although somewhat crude, would have been acutely meaningful to farmers in 1929. Two-wheel drive tractors hitched to a trailing plough were prone to rearing over backwards if their plough became jammed against an immovable object such as a buried stump or rock. Fordson Model F and Fiat Model 702 tractors were particularly vulnerable to this occurrence. This is not a hazard with modern 3-point linkage mounted ploughs.

Opinion

All who have used the FOUR DRIVE TRACTOR have agreed, without a single dissenting opinion, that the FOUR DRIVE is by far the only practical Tractor made and in use today—that the cost of the FOUR DRIVE and the cost for repairs and upkeep, doing the same work of any five-ton Crawler type, and its cost for repairs and upkeep in one year's work, is double that of the FOUR DRIVE—consequently the difference is waste.

We offer the following brief specifications covering THE FOUR DRIVE TRACTOR:

Brief Specifications of Model D
Straight Line Drive

MOTOR	Climax 4 cylinder vertical, 5 x 6½
OILING SYSTEM	Force feed through crank shaft
GOVERNOR	Climax special
FUEL	Kerosene-Gasoline
TANK CAPACITY	28 gallons
DIAMETER BELT PULLEY	14 inches
WIDTH BELT PULLEY	8 inches
WHEEL BASE	7 feet 2 inches
WIDTH	5 feet 9 inches
TOTAL LENGTH	10 feet 10 inches
TOTAL HEIGHTH	6 feet 3 inches
REAR WHEELS	42 inches—12-inch face
FRONT WHEELS	36 inches—12-inch face
TURNING RADIUS	7 feet 6 inches
FRAME	5-inch 9 pounds, steel channel
RADIATOR	9 gallons, Spirex radiating surface 560 square inches
MAGNETO	Eisemann or Dixie with Impulse Starter
CARBURETOR	Kingston or Ensign
APPLICATION OF POWER	All four wheels
HOW APPLIED	Live Axles
REDUCTION GEARS	Nickel steel, heat treated
FRONT DRIVE	Bevel—Brown-Lipe-Chapin
REAR DRIVE	Timkin—Worm and Worm Gear
DIFFERENTIALS	Brown-Lipe-Chapin
AXLES	Chrome Vanadium, heat treated
TRANSMISSION	Cotta; three speeds; direct on intermediate
REDUCTIONS — Reverse	82½ to 1
High	32 to 1; speed 4 miles per hour
Intermediate	50 to 1; speed 2½ miles per hour
Low	80 to 1; speed 1½ miles per hour
TRACTOR RATING	Drawbar 20
BELT RATING	35

Weights

Remember every working part is steel housed and automatically lubricated
IT HAS NO CHAINS OR UNIVERSALS

Model D Equipped Standard for tractor and belt work	6,000 pounds
Model D Equipped with tamping grousers for packing and rolling	10,000 pounds
Model D Equipped with rubber-tired wheels and maintainer	7,500 pounds

Buy a Four-Drive and Save the Waste

The marketing people at the Four Drive Tractor Company could not refrain from inserting some sales text into a page of specifications.

FITCH FACTS  FITCH OWNERS

```
                                        Ponca City, Oklahoma,
                                        October 15, 1919.
Four Drive Tractor Company,
     Big Rapids, Michigan.
Dear Sirs:
        We bought a Four Drive this fall and have been pulling four
14-inch bottom plows.  It pulls two 8' binders, two 8' drills and
a 36' harrow.  We plow from four to ten inches, and it never feazes
the tractor.

        We plowed 75 acres in six days and never rushed things.
It beats anything to pull I ever saw.  The tractor sure has the
advantage over other tractors because of front drive.

        We never have had it stuck since we had it.  The Tractor
is a hog for work, the more you work it the better it likes it.

                                        Yours truly,
                                        David Proctor, R. R. No. 6.
```

CRIGINAL ON FILE

FITCH FACTS  FITCH OWNERS

```
PACIFIC CREAMERY COMPANY
    Lilly Evaporated Milk
       Phoenix, Arizona.                May 18, 1918.

Four Drive Tractor Co.,
Big Rapids, Michigan.
Gentlemen:
        Beg to say that we are very much pleased with the
4-wheel drive tractor which we purchased from your agent, Mr.
Chas. Juncker of Glendale, January 3, 1918. It has done excellent
work; we have pulled three 12" bottom plows in different kinds
of soil from 4" to 8" deep. Was able practically at all times to
pull the plows on intermediate gear. We plowed approximately 100
acres of Bermuda grass sod and disced it all. We find the gasoline
consumption about 12 gallons for 10 hours, and the up-keep of the
machine was a small item.
                                    Yours very truly,
                                    PACIFIC CREAMERY COMPANY.
CWZ:LMF                             C. W. Zimmerman.
```

ORIGINAL ON FILE

Two extracts from the many pages of testimonials contained in a Fitch sales brochure. The reference to 12 gallons for 10 hours is both surprising and impressive.

CHAPTER 10

Buying a Classic Tractor

The following is a lighthearted approach to the formidable task of selecting and buying a classic tractor. The text was originally published in an abridged form in the 'Australian Grain' magazine of July 1996. The reference to dollar values applies to that period.

The difficult decision

Interest in Classic Tractors is escalating at an astonishing rate throughout the length and breadth of our island continent. The enthusiasm is not restricted to ageing farmers but can include retired bank managers, youths with back-to-front caps, school teachers, hobby farmer chaps, Greens and even Volvo owners. A considerable proportion of these are at any one time contemplating their first tractor purchase.

The most common question I am asked is 'What tractor should I buy and how much should I pay?' There is no simple answer to this; however, I have endeavoured to put together some guidelines which may help.

What NOT to buy!

Unless you are (a) a masochist or (b) a gifted mechanical engineer, with the available resources of a well equipped engineering shop and a bank roll to match, my urgent advice is — do not contemplate buying an aged tractor that has been abandoned in the open for countless years. An old timer's sincere estimation of the number of years elapsed since his faithful servant actually 'went' can usually be multiplied by a factor of at least two. Time tends to dim the memory upon these occasions.

Tractors left standing unshedded have an unfortunate habit of becoming waterlogged each time heavy rain descends. A rusty Letona peach tin covering the exhaust pipe is no guarantee that water has not penetrated the inner sanctum of the engine. Frequently, moisture enters around the base of the exhaust pipe where it joins the manifold, or through unprotected air cleaners, down the sides of rusted or removed spark plugs and so forth. Transmissions, steering boxes and magnetos are also highly prone to water entry. Water, of course, creates rust, which is often aggravated by constant evaporation followed by repeated wetting.

Thornbush, lantana, thistles and other assorted noxious nasties entwined around axles and gear levers, or the odd wattle having a love affair with the exhaust pipe, are not encouraging signs. Such adornments suggest that the tractor was in fact not parked there 'just a few weeks ago'.

The message therefore should be perfectly clear. Leave the abandoned rusting ironmongery to the dedicated devotees who have endless time on their hands, coupled with the patience, skill and a rich uncle necessary to transform the impossible into reality.

1932 Massey Harris. This is a good straight tractor with all the body panels in place. The engine can be started with difficulty. A relatively easy restoration. Value around $1500. (Photo: I.M.J.)

The IDEAL purchase

The ideal tractor to buy is one that has been constantly shedded and that *can* be fired into life. Should this involve a fair degree of cranking, coupled to a sprained wrist or broken thumb, do not be dissuaded. Even if you have to tow the cantankerous beast behind the Hi Lux for a kilometre or two before it finally bursts into life, that is O.K. too. The main factor here is that you have been able to get the engine running and have hence been assured that all the pistons, conrods and things are where they should be. You will be content in the knowledge that a screwdriver and shifter should be all that is required to convert the beast into a purring pussy.

Where finances or simply availability dictate that a *non runner* be acquired, this can still be O.K. providing a few precautionary investigative steps are adopted.

1936 International McCormick Deering W30. There is an abundance of W30 tractors available. This is a reflection of their popularity during the 1930s. The unit pictured has been restored at considerable cost and is equipped with excellent tyres. The standard of paint work is not commensurate with the rest of the tractor. A total repaint would be a good investment. Value $2400. (Photo: I.M.J.)

The most important of these is to determine that the engine can be 'turned over' with the crank handle or, in the instance of some tractors, by rotating the flywheel. If the engine is locked tight there is then an unidentified problem. This can range from a major disaster, such as a broken crankshaft, to a relatively minor temporary 'sticking' of the pistons. Frankly, there is no way of accurately diagnosing this apart from having X-ray eyes or instigating a partial dismantling of the engine. If you feel you can trust the vendor (and don't, if he wears white shoes) some light might be thrown on the problem enabling an informed decision to be taken.

Traps for the unwary

Steer clear of tractors that have significant parts missing, unless you are absolutely certain of being able to procure these from some known source. A rare old Sift would be a nothing without its injection pump, a Hart Parr a non-identity without its cast radiator header tank. These crucial components cannot be purchased along with the Sunday papers and a loaf of bread at the corner shop. No matter how complete and desirable the tractor may appear to an intending purchaser, it is really not worth contemplating should it be minus such vital or characteristic essentials.

Dings, splits and rusty gaps in sheet metal components can be difficult and costly to repair or replace. Quite frequently it can be a lot easier and less expensive to perform a mechanical repair than to reinstate a rusted out set of mudguards. Early tractor manufacturers utilised Sydney Harbour Bridge

1936 Fordson N. There are probably more old Fordson tractors around today than any other tractor. Despite this they are gaining favour with collectors. This example is relatively complete and the motor is free. Value $700. (Photo: I.M.J.)

grades of sheet metal. Expect to observe an open-mouthed expression of dismay upon the countenance of a panel beater to whom a pair of Ronaldson Tippett mudguards have been presented for beating out.

Wheel equipment can cost big money

Partially rusted wheels should be contemplated with a jaundiced eye as they often require expensive renovations. A tractor that has been parked for long periods can contract serious rust on the underside of its steel wheels that may not be immediately apparent. In the case of pneumatic tyres, it can easily cost in excess of $1000 to replace a set of perished covers. Additionally, inner tubes today cost the same as did tyres in 'the good old days'.

Price guide

The price of collectible classic tractors in Australia can vary from free to beyond $10 000 for something special. The three determining factors are condition, degree of rarity and the desire of the vendor to sell.

Any tractor that fits the definition of the first type examined in this chapter (i.e. an abandoned, rusted out heap of ironmongery) *should* be value*less*. The cost of simply recovering these machines can be astronomical and could be better spent helping to reduce the National Debt. The thought of hiring heavy cranes, low loaders, etc. is enough to cause palpitations. Seldom, however, is a tractor ever offered free and therefore expect to pay a token price of anything up to $500.

A shedded tractor that is complete, *but not able to be started*, will range from $500 for a relatively common Fordson N to $5000 for an especially rare and very old machine. The same units will attract double these prices if they can be started and driven.

Carefully restored and well presented tractors are much in demand and often enjoy premium prices. Based upon recent sales, the table that follows will serve as a guide to current trends of a cross section of tractor prices.

1944 Massey Harris 102. Mechanically this tractor is in excellent order and all the body panels are remarkably straight, although there is some rust in the mudguards. Value $1800. (Photo: I.M.J.)

Don't rush things

So you have put the new fridge on the back burner and have purchased your pride and joy. Unless the machine has been recently restored and/or *thoroughly serviced,* it is unwise to start and drive the tractor until the engine crank case and transmission housing/s have been drained, flushed and filled with the appropriate oils. The same philosophy applies to the fuel tank, carburettor and radiator. Unpleasant and destructive gooey silt tends to lie in the bottom of these places and is potentially damaging. Grease should be applied liberally to all nipples, making certain that the old grease is displaced.

1918 Bailor. A very rare artifact — possibly only one other in Australia. The tractor runs and is in good mechanical order. Value $13 000. (Photo: J. Rethus)

A final plea

My final comments are issued as a plea. Please remember that your classic tractor is an old lady who has performed worthy service in her younger days. She is now weary and creaking at the joints. Treat her gently and with respect. Don't let the kids race her up and down chasing chooks around the paddock. Restrict her engine revs to little more than half her capabilities. In return — she will last for ever.

A guide to prices

The following is a list of sales that have taken place recently.

1936	McCormick Deering W30	$2000	Restored
1950	Lanz Bulldog Model S	$5500	Good going order
1958	Lanz Bulldog Model R	$6500	Good going order
1949	H.S.C.S.	$5800	Going
1941	Case LA	$3000	Restored
1950	Fordson Major E27N	$2800	Restored
1950	Orenstein & Koppel	$3500	Not going — parts missing
1928	Vickers Aussie	$2000	Not going — rough
1942	Massey Harris 102	$ 800	Going — unrestored
1960	Massey Ferguson 35 diesel	$6200	Restored
1952	Ferguson TE20	$2400	Restored
1952	Farmall Super A	$1600	Restored
1947	John Deere D	$4000	Restored
1932	Massey Harris 25	$3100	Good going order

1960 John Deere Lanz 440. This is a tractor that would attract a great deal of attention if offered for sale. It is particularly interesting as it is the first John Deere produced at the Lanz Bulldog factory in Germany. Only a very few of this model came to Australia. The tow chain suggests that it is not a 'goer', however as the engine is a 3-cylinder Perkins diesel, parts would be readily available. Value $2200 as is. (Photo: G. Baker)

CHAPTER 11

Kubota — The Late-rising Sun

Kubota rates as a major player in the contemporary world tractor scene. Considering the first tractor to bear the name was introduced as recently as 1960, this is an extraordinary achievement. Its 'late arrival', however, renders the Kubota ineligible to be deemed *classic*. Despite this, as little has been written about the Kubota tractor story, it is worthy of a brief mention within these pages.

The Kubota dawn

In the year 1890 Gonshiro Kubota established an iron casting foundry in Osaka. Within three years the firm's engineers had perfected a technique for producing cast iron pipes. This represented a major breakthrough for Japanese municipal authorities, who hitherto had relied principally upon bamboo for ducting urban water supplies.

Kubota's first involvement with precision machinery was in 1897, when it acquired a match-manufacturing company. Twenty years later, in 1917, Kubota attracted world attention when it unveiled an advanced design for marine steam propulsion.

But it was not until 1947 that a restructured Kubota became involved with mechanised agriculture. Like so many nations at that time, Japan's population began a period of rapid escalation. The demand for increased food production was pressing. Rice was the main crop, but traditional methods of cultivation based on hand labour and primitive tools could not cope with the challenge. Accordingly, Kubota introduced an engine-driven, walk-behind cultivator that could operate tirelessly in the ankle deep water of the paddy fields. One man and a cultivator could achieve in a day the work of twenty men with hand tools.

Following years of exhaustive testing, the first conventional Kubota tractor was released to Japanese farmers in 1960. Built in the Naniwa-Ku works at Osaka, it was a unit frame construction modern lightweight tractor, built along similar lines to its European counterparts such as Eicher and Kramer. The 15 h.p. tractor was designated the T15 and painted red with white highlights and black mudguards. An 18 h.p. version, the T18, soon followed. The T series was superseded by the L series, which ranged from 12 to 27 h.p. The bodywork of the L series was painted yellow with grey engine and transmission and red wheels. Later L series tractors were coloured Kubota orange with grey/blue engines and transmissions. This was considered an ideal colour scheme and has been carried forward into the current series.

In 1971 Kubota sent a tractor to Nebraska for testing at the Tractor Test Laboratory. The Model L210, weighing 2320 lbs, was powered by Kubota's own 2-cylinder liquid-cooled 4-stroke vertical 63 cu. inch diesel engine. It produced 19.1 p.t.o. h.p. at 2700 r.p.m. The tractor was comprehensively equipped with a full lighting kit, 3-point linkage, a hydraulic remote outlet, p.t.o. shaft plus a 6-forward and 2-reverse-speed transmission.

The T15, introduced in 1960, was the first Kubota conventional tractor. (Photo kindly sourced from Japan by Kubota Tractor (Australia) Pty. Ltd.)

KUBOTA MODEL L210

1971 Kubota Model L210. Its twin-cylinder diesel engine had indirect injection combustion chambers and required glow plug pre-heating prior to starting. (Line drawing from original Kubota promotional material)

The Model L210 was an instant success both in Japan and the United States. Other types were to quickly follow and solid export markets were established throughout Asia and beyond. The first Kubota tractors to enter the Australian arena arrived in 1974. They were distributed in the eastern mainland States by Mobilco Co. Ltd., in Tasmania by F & G Webster Pty. Ltd., and in Western Australia by Fred Hopkins Co.

Kubota today

From a late beginning, Kubota now claims to be 'the world's leading producer of mid-size equipment for mechanised agriculture'. Whilst this pronouncement no doubt attracts some comment from opposition tractor companies, there is no disputing that within a few years of obtaining a toehold in any specific area, the under 40 h.p. class of tractors is dominated by Kubota. This is evident in Australia, where for the past decade Kubota has been a clear sales winner in the small tractor class.

Although not yet generally recognised as a major competitor in the larger horsepower class, the progress of the new Kubota 80 Series is being carefully monitored by its competitors. Ranging between 77 and 100 h.p. these good looking tractors are powered by a 4-cylinder diesel engine coupled to a 24-speed fully synchronised gearbox with shuttle directional control.

The Kubota M8580 4-wheel drive is offered with either a ROPS frame or (as pictured) a stylish, functional, fully enclosed cabin. Note the high clearance and the exceptional visibility for the operator. (Photo courtesy Chelmsford Farm Machinery, Taree, N.S.W.)

The Kubota 80 Series tractors are manufactured at the firm's Tsukuba plant. This is the world's largest factory engaged exclusively in the production of farm tractors. Being an integrated production facility, practically all components are actually made within its walls. The entire production is controlled by a computer-automated centralised system.

Export sales throughout the world of Kubota farm equipment and engines in 1995 totalled US$1147 million — an increase of 1.7% over the previous year, despite a global recession in farm incomes. With declared assets of US$14 176 million (Kubota Corporation and Consolidated Subsidiaries 1995 Annual Report) Kubota has the financial muscle to further expand its share of the world tractor market should it so choose.

Principal Achievements in Kubota's 105-Year History

- **1890** Kubota begins operations as a castings manufacturer under the direction of its founder, Gonshiro Kubota.
- **1893** Kubota commences the production of cast iron pipe.
- **1897** Kubota takes over a company that makes machinery for the manufacture of matches, thus laying the foundation for the Company's machinery operations.
- **1917** The Company commences the manufacture of steam engines for ships.
- **1936** Kubota succeeds in industrializing its centrifugally cast iron pipe.
- **1947** The manufacture of power tillers begins. This later becomes the foundation of Kubota's farm equipment operations.
- **1952** Kubota begins the full-scale production of pumps to expand its business in water-related markets.
- **1956** Kubota begins the manufacture of large-scale construction machinery.
- **1960** Kubota commences the manufacture and marketing of colored cement roofing materials.
  Kubota develops the first Japanese-produced tractor.
- **1962** The Company's water treatment business begins.
  The marketing of centrifugally cast steel G-columns, a structural material for construction use, begins.
- **1964** Kubota begins the production of ductile tunnel segments.
- **1977** Fire-resistant siding materials are developed.
- **1985** Kubota enters the computer and computer-peripheral markets.
- **1986** The Company commercializes its melting technologies for waste material treatment.

CHAPTER 12

David Brown — The Innovator

In the year 1932 David Brown II assumed control of the company David Brown Ltd. Founded by his grandfather in 1860, the firm had grown from a three-man team to become Britain's largest manufacturer of industrial automotive and marine gears.

A 'gentleman farmer' and his tractors

In addition to the demanding duties of his new appointment as chairman, David Brown II was also a 'gentleman farmer' and developed a keen interest in farm mechanisation. He became fascinated with the experimental work being carried out by his friend Harry Ferguson in Northern Ireland. This concerned the design of a light tractor incorporating Ferguson's patented 3-point linkage implement attachment system and weight transfer control. Brown shared the Ulsterman's vision that the concept would eventually be adopted by tractor manufacturers around the world — a prediction that was to be proved accurate (see 'Ferguson — What *Really* Happened', Chapter 19).

Following negotiations, a special plant was constructed within the David Brown works, located near Huddersfield, Yorkshire, to produce Ferguson's tractor. The project was financially backed by David Brown II personally and operated independently from David Brown Ltd. A new company, David Brown Tractors Ltd., was registered.

The first David Brown-made Ferguson Type A was released to farmers in 1936. Sales were disappointing, largely due to the depressed state of the British rural economy at that time.

The business relationship between Brown and Ferguson was dissolved in 1939, with Ferguson taking his design over the Atlantic to Henry Ford.

In the meantime David Brown II had toured the British countryside seeking the opinions of farmers about *their ideas of what constituted the ideal tractor*. As a direct result of his fact-finding travels, the David Brown VAK 1 tractor was launched just weeks prior to the outbreak of World War II in 1939. The VAK 1 was the first of a range of well engineered modern tractors produced by David Brown Tractors Ltd. and marketed under the David Brown banner.

By 1956 David Brown Tractors Ltd. had become the third largest manufacturer of agricultural tractors in Britain. These gaily coloured wheeled and crawler tractors, painted in Hunting Pink, earned valuable export income for a country whose economy was still recovering from the ravages of the war. During that year (1956) a new model of significant importance was introduced into the already comprehensive David Brown range. It was designated the 2D.

Why the engine at the front?

For some time David Brown design engineers had unknowingly been working along parallel lines to that of a number of European tractor designers. The 'think-tank boffins' of the German firms of Eicher, Fahr, Fendt, Lanz, Wesseler, and the East German Geratetrager, were all asking the same question: 'Why is the engine of a farm tractor located at *the front* of the tractor?'

Every tractor driver is aware of the difficulties experienced with the precision required to navigate along row crops, with a cultivator attached behind, set up to work centimetres from the crop. It takes considerable skill and steadfast concentration if one is to avoid eliminating valuable seedlings. It is necessary to peer around the front bonnet to maintain accurate steering whilst managing to conduct a constant vigil of the rear implement. This requires repetitive twisting of the neck and shoulders, frequently leading to muscular and spinal problems.

1950 David Brown Diesel Cropmaster. Its popularity and resulting sales were responsible for elevating David Brown into being one of Britain's major tractor manufacturers alongside Ferguson and Fordson. The tractor pictured is part of a David Brown collection owned by a farmer in the windswept Orkney Islands. There would be few tractor collections closer to the North Pole than this! (Photo: I.M.J.)

A 1930 Fendt one-cylinder 6 h.p. diesel tractor with mid-mounted sickle mower and rear single-furrow mouldboard plough. Despite its archaic appearance, records indicate that the machine performed efficiently. (Photo courtesy Fendt, Marktoberdorf)

During the 1930s Fendt experienced a decade of steady growth and development. On 31 December 1937 the restructured firm of Xaver Fendt & Co. moved into its newly completed modern plant at Marktoberdorf. To mark the occasion, a new model Fendt Diesel Horse was added to the expanding range of tractors. This was the F18 and was significant for the fact that it was the first European tractor to be fitted with an independent power take-off shaft that could be engaged under load. (It is interesting to note that the first North American tractors to incorporate this principal were the Cockshutt and Sheppard tractors, over a decade later.)

By late 1939 the output from the Fendt Marktoberdorf plant had expanded to over 1000 tractors per annum.

The war years

During World War II it became mandatory for all German farm tractors to have their engines modified to run on wood gas, or remain in the barn. Accordingly, many tractors were equipped with charcoal/wood burners. These cumbersome add-ons produced a gas of low but adequate volatility.

Fendt released its *purpose built* gas burner tractor in 1942. Known as the Holzgasschlepper Type G25, it was powered by a custombuilt Deutz Gasmotor which developed 25 h.p. The tractor performed reasonably efficiently, despite the fact that most gas producers fitted to tractors (and indeed cars, trucks and buses) were not noted for their reliability. The gas producers served purely as a stop-gap measurement during the critical fuel shortages of the war years and immediately after.

A 1942 Fendt Holzgasschlepper Type G25 with a 2-cylinder Deutz Gasmotor 25 h.p. engine. Fendt was one of an elite group of tractor manufacturers who took the initiative to construct a purpose-built wood burning gas producing tractor. Others were Fahr, Hanomag, Normag, Deutz, Schluter, Lanz and Famo. (Photo courtesy Fendt, Marktoberdorf)

Post-war development

By 1948 the Fendt factory was again in full production, but now under the control of the Fendt brothers Hermann, Paul and Xaver. They were obliged to compete in their home market with no fewer than 50 opposition tractor manufacturers. Despite this, the Diesel Horse range was extended and by 1949 comprised 20 different models.

An important development took place in 1953 with the introduction of the Fendt Geratetrager Type F12 G.T. This consisted of a rear-engined tractor (or implement frame) which included a removable front transport box. It featured four implement mounting positions — front, rear and two underbelly. The philosophy behind the design was that the operator could achieve more than one work application with only a single pass. Additionally, the operator enjoyed an unobstructed view, enabling extreme accuracy during row crop cultivation.

1953 Fendt Geratetrager Type F12 G.T. This 12 h.p. multi-purpose tractor was ahead of its time. Fendt introduced a second version in 1965, this time with a 30 h.p. engine. It was not until the 1990s that farmers finally appreciated the concept of a multi-implement tractor having its engine mid- or rear-located. (Photo courtesy Fendt, Marktoberdorf)

The Geratetrager represented a bold concept ahead of its time. Other manufacturers experimented with the same principal. Allis Chalmers had its M, Lanz its Alldog and David Brown its 2D, and there were several other German manufacturers exploring similar design ideas (see 'David Brown The Innovator', Chapter 12.) All of these were conceived in an era when rear-mounted 3-point linkage implements were being promoted to farmers throughout the world. As a result the Fendt Geratetrager, along with the other rear-engined tractors, faded from the scene after only a few short years.

The 100 000th Fendt tractor came off the assembly line in 1961. It was a 30 h.p. Farmer 2, and was one of the most technically advanced and aesthetically pleasing tractors of the period.

A 34 h.p. 1961 Fendt Farmer 2. During the 1950s Fendt placed a great deal of importance upon the stylish appearance of their tractors and the attractively styled Farmer 2 was evidence of this. However the tractor not only looked good, it was also extremely functional and incorporated the latest in tractor technology. (Photo courtesy Fendt, Marktoberdorf)

Fendt — heading for the 21st century

As in the days of Sylvester Fendt the clockmaker, the company continues to enjoy a reputation for producing soundly engineered products that have been thoroughly researched and tested prior to their release.

During the 1970s and '80s Fendt achieved some notable accomplishments, thus consolidating its grip on the European tractor market:

1970 Fendt acquired the enterprise of Lely Dechentreiterin, a company well respected for its range of tractor implements and machinery.

1976 The introduction of the 6-cylinder Favorit LS range (85-105 h.p.) incorporating a 40 k.h.p. transport speed. This high-speed gearing was a major breakthrough welcomed by European farmers, who often travelled considerable distances on public roads between farms. Other tractor producers quickly followed the Fendt example.

1985 Fendt succeeded for the first time in gaining number one position in the official records of tractors registered in Germany.

1988 Fendt became the top-selling tractor in Holland.

1989 Fendt was awarded a gold medal at the prestigious SIMA Paris Show for its unique design of Swinging Power Lift. This clever engineering initiative allowed 3-point linkage mounted implements to be adapted for working on difficult slopes. Both lift and lateral swing could be hydraulically controlled from within the cab.

In 1994 Xaver Fendt GmbH & Co. produced an impressive 7370 tractor units which generated a turnover of 699 million marks. In terms of world tractor production the company is a relatively small player compared with the vast output from the new conglomerates such as New Holland (Fiat-Ford), Case International, Agco (Allis-White-Massey Ferguson) etc. It is therefore commendable that the results of the Fendt research and development team enable the company to continually release new models that are at the absolute cutting edge of tractor technology.

A further example of this Fendt technology is the Xylon Series, which could be construed as the modern equivalent of the innovative Geratetrager, introduced back in 1953. The Xylon models are powered by a range of M.A.N. diesel engines ranging from 110 to 140 h.p. placed on their sides under the centre of the tractor. This permits the forward, mid and rear mounting of implements, all of which may be operated simultaneously during one pass. The Turbo Shift gearbox provides a top road speed of 50 k.p.h. The low centre of gravity and the powerful 4-wheel brakes enable the Xylon to travel at this high speed with a wide margin of safety.

On 18 July 1996 the Board of Fendt issued a press release announcing that the 500 000th Fendt tractor had been produced on that day. This was exactly 68 years after Johann and Hermann Fendt had proudly exhibited their first tractor to a group of fascinated Bavarian farmers.

This 1996 photo of three Fendt Xylon tractors graphically demonstrates the multi-implement tractor concept, which was originally trialled with the Geratetrager in the 1950s and is now being enthusiastically embraced by European farmers. Australian and North American farmers have been slower to appreciate the labour- and fuel-saving advantages of the design. However, attendances at recent field days indicate a rapidly growing interest. (Photo courtesy Fendt, Australia)

CHAPTER 14

The Mail Order Graham Bradley

Buying from mail order catalogues was a way of life for country folk who lived some distance from major commercial centres. The practice was common in Australia, Canada, the U.S.A. and the more remote regions of the British Isles. Catalogues offered a varied range of products, equivalent to what might be expected in a well stocked city department store.

Sears Roebuck

The most renowned and comprehensive of all mail order catalogues were those that sent out to homesteaders in the U.S.A. by Sears, Roebuck & Co. These bulky catalogues are now considered collector's items and constitute a concise historical record of merchandise of days long gone.

The range of products offered by Sears Roebuck was staggering. A pin to an anchor or a cotton reel to a wedding dress is not an exaggeration. Even two-storey homesteads could be ordered off the catalogue. They were delivered to the nearest rail head and only had to be 'nailed together'. The company also did a brisk trade in every conceivable type of farm gadgetry and machinery. Indeed manufacturers were often sorely pressed to meet the demands of supply necessary to fulfil the Sears Roebuck orders. As a consequence, the mail order firm created its own brand of farm machinery which was named 'David Bradley'.

In the mid-1930s, encouraged by the volume of orders pouring in for David Bradley hay and tillage equipment, Sears Roebuck joined forces with the Graham Paige Motor Co. of Detroit and put together a team of young, visionary design engineers to create a modern state-of-the-art tractor. Graham Paige agreed to produce the tractor equipped with its own design 6-cylinder side valve 199 cu. inch engine.

The mail order tractor

The Graham Bradley tractor was released to farmers through the Sears Roebuck catalogue in 1937. Its introduction confounded the established traditional tractor companies. Here was a mail order firm and a car company who in two years could design and have in production a tractor that was broadly equal to the best the others could offer after years of tractor manufacturing experience! There was no doubt the Graham Bradley was a splended tractor brimming with desirable features.

Whilst the efficiency of a tractor should not be judged by its appearance, the mail order tractor was a superbly styled machine. Certainly 'beauty is in the eye of the beholder', but few would disagree that the appearance of the Graham Bradley eclipsed the styling being produced by Oliver, Massey Harris, Case, etc. But its beauty was not just skin deep.

The 6-cylinder engine was governed to a relaxed 1400 r.p.m. and produced 32 brake h.p., which was sufficient to easily handle 2 x 16 inch or 3 x 12 inch mouldboard ploughs. The 2nd gear speed of 4.4 m.p.h. translated to 15 acres of ploughing in an 8-hour day. (This may seem insignificant compared to the cultivation that can be achieved with broadacre equipment pulled behind modern high horsepower tractors. It should be remembered, however, that the figure of 15 acres related to mouldboard ploughing a mere 3 feet width strip on each pass.)

The Graham Bradley had four forward gears, including a transport speed of 22 m.p.h. Only a light clutch pedal pressure was required to operate the single-plate semi centrifugal Velvet-grip clutch. Drum brakes could be used individually for turning or jointly for road work. A centre p.t.o. shaft, swinging drawbar, lighting equipment and comprehensive instrumentation were all standard equipment. The side mounted belt pulley was driven from the rear of the gearbox and had *4 forward and 1 reverse speeds*. This was a clever innovation which could result in considerable fuel savings by adjusting engine r.p.m. and pulley gearing to suit individual applications.

A 1938 Graham Bradley owned by Vern and Grace Anderson of Lincoln, Nebraska. This is the row crop tricycle version which was particularly popular with farmers who predominantly grew corn. Note the intricate steel casting of the front grill and the totally enclosed bonnet line. (Photo: I.M.J.)

The overall impression of driving a Graham Bradley was one of silky smoothness. Everything seemed to happen without fuss. All the controls were light, including the steering, which was of a design used by some earlier tractors, including the Wallis Bear. It featured a drag link attached to a roller chain wrapped around a circular grooved casting fixed to the front axle by a vertical spindle.

Sales incentives

Today's agriculturalists might find it incomprehensible that farmers of the 1930s would even contemplate purchasing a tractor through a mail order firm. A contemporary farmer need only flick a few keys on his FAX machine to attract the attention of an enthusiastic tractor salesman, but back in the pre-war days many of the farms on the North American prairies were remote from tractor dealers. Such was the acceptance of Sears Roebuck as a reliable provider of quality goods that few farmers would have harboured any hesitation in posting off a mail order accompanied by a cheque for the deposit.

Graham Bradley tractors sold particularly well in Iowa, Indiana and Illinois. By overriding the governor, the streamlined machine could be driven at speeds up to 40 m.p.h. pulling a wagon. As the fee for a tractor road permit in these States was only $8, compared to around $60 for a truck permit, many farmers sold their conventional tractor plus their truck and ordered a Graham Bradley.

From 1938 Graham Bradley tractors were equipped with a bigger capacity engine. It was a Continental 217 cu. inch 6-cylinder side valve unit. Whilst producing similar horsepower to the Graham Paige engine it had a greater torque backup, thus enabling the tractor to 'hang on' longer in a specific gear when encountering a stiff patch of soil or an incline.

Illustrated is Vern and Grace Anderson's 1938 Graham Bradley with the bonnet side cover removed to show the 6-cylinder Continental side valve 217 cu. inch motor. (Photo: I.M.J.)

Contributing to the popularity of the Graham Bradley was the policy of Sears Roebuck in offering deals that no opposition company could match. Upon receipt of a 10% deposit, the mail order company would land a brand new tractor on a farm whilst giving the farmer five years to pay off the balance at an attractive low rate of interest. Further, a new plough *and* cultivator would be 'thrown in' at no extra cost. These were compelling reasons for a farmer to bypass the traditional tractors, particularly as the delivered price was around the same as for a John Deere or Allis Chalmers of the equivalent horsepower.

The intervention of the war in 1941 and the priority requirement for Graham Paige to switch to military production marked the end for this remarkable tractor, following the production of 1596 units. The marque was not revived after the war.

Guaranteed - BY SEARS

The GRAHAM-BRADLEY had to make good - had to outshine every other tractor of its size in power and economy - before Sears would offer it to the American farmer; before they would endorse it with the famous Sears money-back guarantee, the fairest guarantee ever written to cover a farm tractor.

The GRAHAM-BRADLEY exceeds our greatest expectations and is now ready for the American farmer. Sears announce this great new addition to their farm machinery line with pride and confidence!

A sample of Sears Roebuck advertising in 1937. The Graham Bradley was the only tractor in the U.S.A. to be sold with an explicit money-back guarantee.

CHAPTER 15

Ludwig Simon

I am indebted to Hermann Simon for permission to recount my personal experiences involving his late father, who I was privileged to know as a friend and colleague. My appreciation is also extended to my old friend Ulrich Schultz for pointing my memory in the right direction.

My first encounter with Ludwig Simon took place in 1954. I was a 19-year-old Scot gaining experience in Australia, which was intended to benefit and prepare me for my future in the world of agriculture. At that time I was employed driving a KL Bulldog tractor for a share farmer. A few weeks previously I had extricated myself in one piece from doing a stint as a windmill erector's assistant, and this had been preceded by a brief but earthy education as a roustabout for a shearing contractor. Wet weather had temporarily halted all ploughing activities in the Narrabri area so I had taken the opportunity of journeying to Sydney, having promised the sharefarmer I would call at the premises of Lanz Australia Pty Ltd to pick up some spare parts for the Bulldog.

The initiation

Lanz Australia Pty. Ltd. had not yet moved into its newly renovated premises at Surry Hills and still occupied the first floor of an office complex at 35 Pitt Street. As I neared the top of the narrow flight of stairs I encountered a tall, imposing, aristocratic German Count glaring down at me. Not that he was wearing his purple sash and Imperial Star, but I had seen all the movies and he couldn't fool me! But to be honest, he only resembled a Count. I had in fact, for the first time, come face to face with Ludwig Simon, Chief Engineer of Lanz Australia Pty Ltd, certainly the most brilliant tractor technician I would ever be privileged to know in my long and continuing association with the tractor industry.

'Vot is it you vont?' he barked at me in a heavily accented parade ground voice. I explained I was in search of some spare parts for a Bulldog, but when I added ' ....a KL Bulldog' I could feel the somewhat chill atmosphere dropping quite a few degrees.

I was to find out repeatedly in the coming months that the Germans considered the KL Bulldog to be quite inferior to the Mannheim original.

However, they took relish from the fact that the Australian Bulldogs were out of production and once again the 'pure Fatherland superior *Lanz Bulldogs*' were being offered to Australian farmers. Anyway, Ludwig Simon assured me that *despite* the parts being for a KL, he could indeed assist me and left me in the care of Ulrich Schultz, the spare parts manager.

When I was ready to depart, the severe erect figure of Ludwig returned accompanied by a large gentleman with a booming accented voice, horn-rimmed glasses and a pipe. He was introduced to me as Herr Tronser, the Managing Director. I was ushered into a large office and invited to accept a seat. I remain bewildered when I try to remember what exactly followed, but I recall that the two gentlemen interrogated me at length about my jackarooing experiences and my earlier teenage years on the farms in Scotland.

All this was rather overwhelming for a young British lad whose formative years had been spent during World War II learning to fear the spread of the Third Reich. Ten years previously we had still been at war with them! This was my first personal encounter with the enemy.

Herr Tronser was no shouting strutting interrogator. Admittedly he had a loud voice but it was tempered with kindness and indeed compassion, particularly when he learned that I had journeyed to Australia on my own. Could this kindly pipe-smoking gentlemen who, apart from his accent, might have been a concerned uncle, really be a German?

Ludwig Simon, Chief Engineer of Lanz Australia Pty Ltd. (Photo: I.M.J., 1957)

'Herr Tronser studied me from across the desk.' (Photo courtesy John Deere-Lanz Archives)

So I am not sure how it happened, but I do know that upon my return to the normality of the footpath in Pitt Street I was no longer a tractor driver for the share farmer. I had accepted a position with Lanz as a field representative, due to commence my new job in one month's time.

'It is not right!'

I commenced work in early 1955 at the relocated Lanz premises, which were situated just a short distance from Central Railway Station at 464 Bourke Street, Surry Hills, in a narrow-fronted converted block of four terrace houses. Downstairs was the warehouse where the Bulldogs were unpacked from their crates. Upstairs consisted of offices at the front and a spare parts store at the rear.

When the Bulldogs were started up downstairs (if you follow me) for testing, the girls in the office upstairs had to immediately grab pencils and pens in a reflex action to prevent them vibrating off their desks. The whole building shook! Herr Tronser frequently emerged coughing and gasping from his glassed-in office as clouds of black exhaust smoke belched through the floorboards from the workshop below, causing his executive carpet to almost float.

The shaking building created havoc with the telephone switchboard and phones jangled everywhere, adding to the commotion. Structural engineers were rushed in to reinforce the upstairs floor in case the weight of the spare parts, coupled to the vibrations from the tractors, caused the whole place to collapse.

The Lanz premises at 464 Bourke Street with Bonkowski's service van parked outside. Herr Tronser's office window is immediately above the parked tractor. (Photo: I.M.J., 1956)

One morning as this usual bedlam was taking place, I was seated across the desk from Ludwig Simon, when he paused in mid-sentence and cocked his ear to the sound of a Model R Bulldog doing its best to loosen the foundations. He looked at me with an expression of annoyance, then stated 'It is not right! Why do they not correct it?' Apparently his stereophonic ears had detected something amiss with the tractor. To mere mortals the sound seemed identical to that which we had grown to expect several times each day. However Ludwig was no mere mortal, as I was to learn

He grabbed the phone. 'Tell Bonkowski to come here. At once!' he shouted.

The tractor was mercifully throttled back and almost immediately the service engineer Arthur Bonkowski, who had been directed to Sydney from

Shanghai, arrived at the top of the stairs and entered Ludwig's office. Having by now been with Lanz for some weeks, I no longer marvelled at the way the German mechanics exhibited a demeanour of humble deference when in the presence of their 'superiors'. Bonkowski doffed his denim cap and stood rigidly to attention awaiting his orders.

'You must first remove and clean the atomiser. There is a blockage in one of the orifices.'

Bonkowski bowed his acknowledgement of the directive, clicked his heels and dispatched himself downstairs.

Twenty minutes later the Model R was re-started. It sounded precisely the same to me, but Ludwig nodded with knowing approval. This was the first of many insights I had into the genius of this extraordinary man.

The Leeton Service School

By 1957 the population of the Lanz Bulldogs was rapidly expanding in the Murrumbidgee Irrigation Area. John Ross-Reid, who was my local dealer based at Leeton, discussed with me the possibility of staging a Lanz Bulldog service school in the interests of the many new owners. I agreed, so John negotiated the rental of a pavilion at the Leeton Show Ground. I persuaded Ludwig Simon to honour the district with his presence and conduct the three-hour evening course. Bonkowski was instructed to be in attendance, thus protecting Ludwig from the distasteful chore of having to soil his hands on greasy components.

'John Ross-Reid was my local dealer at Leeton.' (Photo: L. Simon, showing John Ross-Reid demonstrating a Bulldog Model H, to which is attached a potato planter made by Port Implements of South Australia)

It was a cool autumn evening. About 50 farmers arrived in dribs and drabs at around 6 p.m. The pavilion buzzed with their conversation. The promise of ale and sausage rolls at the conclusion of the school had no doubt contributed to the excellent roll-up. A particularly old and well worn Model P Bulldog had been borrowed for the occasion. Ludwig had expressly stated that, as certain adjustments would be demonstrated, he wanted a tractor that *required* adjusting.

A Model P Lanz Bulldog similar to that requested by Ludwig Simon for his Leeton Service School.
(Photo: I.M.J., courtesy A. Latimore of Comboyne, N.S.W. The driver is Mathew Latimore)

Ludwig Simon, resplendent in a starched and pristine white dustcoat, stood on a dais that had been hastily constructed for him. We were well aware that a Teutonic dignitary of such stature as Ludwig Simon should rightfully be elevated in the presence of common people.

There he stood, tall, erect, immaculate, with the stern expression upon his countenance of a Wermacht colonel about to address a parade of delinquent foot soldiers.

The farmers occupied only the rearmost rows of seats and guffawed and joked with one another, oblivious of Ludwig's readiness to commence the proceedings.

I coughed and mumbled something aout 'a little shoosh'. It was obviously inaudible, as the yarning continued unabated. In a well projected voice this time I appealed for 'a bit of quiet'. Still the farmers continued on with one another about the government, irrigation taxes and the weather. Ross-Reid failed to set an example as he yattered away with the noisy Williams brothers from Barellan.

I could see Ludwig was not amused. He had not flown up from Sydney in the Butlers Airline DC3 to be ignored. In a voice that would have rattled windows back in Wagga Wagga, Ludwig erupted. 'THERE WILL BE UTTER SILENCE!' ... and there was! Instantly everyone jumped and looked round at Ludwig in shock. That is, except for poor old Bluey Brown, who really was quite hard of hearing. He continued in a shouting, protesting tone about the cost of fertiliser for about half a minute before realising that the bedlam had ceased and his was now the only voice that disturbed the almost tangible silence. Bluey stopped in mid flight and gazed wildly around, only to be confronted with a withering frown from Ludwig Simon. Order was now restored.

I introduced Ludwig, using the impressive full title, Chief Engineer of Lanz Australia Pty Ltd, which I knew he would have expected. Ludwig bowed to the assembly, much in the manner of a Roman Emperor acknowledging the adulations of his hordes — then frowned as nobody cheered

'We will first talk about the Lanz clutch' he began in clipped martial tones, heavy with German accent. 'I want anyone who does not understand the principal of the Lanz clutch to stand.'

Nobody stood. Furtive glances were flashed around. 'I will say again. Anyone who does not understand the workings of the Lanz clutch will stand!'

Still not a movement. 'So! You all understand about the Lanz clutch, eh? We will see!' he added contemplatively. His gleaming glasses roamed the rows of seats.

Then like a gunshot 'YOU! You will arise and tell us about the Lanz clutch.' His outstretched finger pointed accusingly at a squirming Bluey Brown. Retribution time, I thought.

'Bbbbut I don't know anything about the Lanz clutch', pleaded poor Bluey, appalled at again being the centre of attraction.

'So! I wonder who else does not know about the Lanz clutch?' Ludwig murmured as once again his gimlet eyes roamed the now utterly cowed audience.

'YOU!' Again he caused everyone, including myself, to jump. This time the terrible finger of accusation was pointed at Farty MacFadden.

The shocked Farty blurted out 'Who, me?'

'Yes. You!'

'Stone the crows! I wouldn't have a clue', gasped Farty as he looked around the gathering for support.

'You mean, you also do not know about the clutch?' exploded Ludwig with raised eyebrows. 'So! I will say one more time', now in a quiet, menacing tone. Then almost in a whisper 'You will stand if you do not understand the principal of the Lanz clutch' he hissed.

Instantly there was pandemonium as, to a man, everyone sprang to his feet. Chairs were knocked over and some fearful farmers even waved their hands in an anxious desire to exhibit clearly there lack of knowledge of the Lanz clutch.

Ludwig smiled and nodded 'That is better. Now we can commence the school.'

In fact, the evening was a great success. The farmers soon relaxed when they realised that Ludwig had probably only deployed psychology to gain their undivided attention. But they were not *quite* sure, and certainly Ludwig did obtain their full concentration during the entire evening. The farmers gained an immense amount of useful knowledge. Ludwig was undoubtedly an accomplished tutor and the farmers quickly came to realise and appreciate this fact.

The school concluded at around 9 p.m. and we all adjourned to the Leeton Soldiers' Club, where a room had been set aside. Refreshments were served courtesy of John Ross-Reid. The farmers crowded around Ludwig. They were now hungry for more Lanz information and plied him with questions.

When asked about the damage sustained by the factory during the final days of the war, Ludwig turned quite melancholy and his eyes glassed over. 'It was not right', he intoned sadly. 'The British and American bombers came over every day.'

Bill Williams, a Bulldog owner and an ex R.A.A.F. pilot, had been attached to an R.A.F. Bomber Command squadron in Britain, during the war. This fact was known to everyone in the room apart from Ludwig, who complained on about the rough treatment Germany had experienced from the Allied bombers.

'WHAT ABOUT COVENTRY?' For the second time that evening the silence was absolute. This time it was Bill Williams who had dramatically taken centre stage. Realisation struck Ludwig. He immediately mentally summed up the situation and knew he was on dangerous ground. He also no doubt remembered that he was in a clubhouse built in honour of Australian fallen and returned service men and women.

'I think', Ludwig said slowly, 'maybe it is better I, how you say, shout the drinks and we forget about the past'.

I breathed a sigh of relief. John Ross-Reid caught my eye and gazed heavenward as he mouthed a silent prayer of thanks. The entire group then relaxed.

Two hours later we were poured out of the club by the kindly and understanding doorman. The farmers jostled with each other to put their arms in a brotherly manner around Ludwig's shoulders, with emotional requests for him to return to Leeton and warm invitations to visit their farms. He was now their mate!

Ludwig in India

Prior to his arrival in Australia, Ludwig Simon spent some time for Lanz in India. His brief was to resurrect Bulldog sales following their absence during the war years. This involved the appointment of new dealers, the staging of field days and the courting of high government officials, without which business could not be conducted in India. The following photos were taken in 1950 and present a graphic glimpse of Ludwig's trials and tribulations during his term on the Sub Continent.

A crated Bulldog (upside down despite the arrows) being loaded onto a camel cart at the Karachi wharfs. Note the steam crane. (Photo courtesy H. Simon)

The camel, now with its cart loaded with crated Bulldog components, gazes down disdainfully upon an assembled product. (Photo courtesy H. Simon)

Ludwig puts on a brave face as he conducts a training school at the premises of the Hundustan Electric Co. of Bombay. Note the spilled fuel. Was it melted down camel fat they used? (Photo courtesy H. Simon)

The group on the tractor, with Big White Hunter Ludwig at the wheel, would have been more appropriately mounted on an elephant as they swept majestically through a native village – much to the excitement of the inhabitants. (Photo – courtesy H. Simon.)

Sabu Singh proudly poses beside his ageing Bulldog. 'It is running all through the war you know, with the beautiful fat skimmed from the grease traps of the very extremely best hotels', he explained triumphantly to Ludwig. By the look of his trousers, no one would argue with him. (Photo courtesy H. Simon)

On 29 July 1950 Ludwig arranged the demonstration of a steel wheeled Model L, on a farm near Delhi, to no less than the Indian Minister of Supply and Industry. It was necessary for the tractor be 'trialled' by a soldier prior to the demonstration, in order to determine the suitability of a Bulldog to be gazed upon by such an eminent government minister. Following a brief opening of the furrow by Ludwig, the tractor was handed over to the soldier. (Photo courtesy H. Simon)

In the afternoon of the same day, the Honourable Shree Harekrusha Mahtab and his entourage arrived at the site. Much to Ludwig's consternation, the exalted Minister insisted upon having a drive! The photo shows the Minister wrestling with the controls, totally unaware that the tractor has become bogged (note its angle) whilst Ludwig, balancing on the rear, looks on anxiously but cannot get to the clutch because of the Minister's youthful assistant standing in the way, smiling at the camera. The skirted gentleman on the extreme left appears unamused by the entire spectacle. (Photo courtesy H. Simon)

During a demonstration staged by Ludwig in the arid Jodhpur region, a farmer who was given a drive of the tractor panicked as the tractor and plough came close to the ditch. He jumped off whilst the tractor was in motion and ran off into the bush, leaving the tractor to find its own way into the ditch. The spudded rear wheels continued to rotate, as if endeavouring to bury the tractor. (Photo courtesy H. Simon)

CHAPTER 16

The Aultman Taylor 30-60

Reflecting back over a century of tractor production it is interesting to contemplate a few statistics. For instance, how many different manufacturers (past and present) have existed world wide? The question has little merit as it is simply not possible to arrive at anything like an accurate figure. It is known, however, that in North America the total exceeds 900, if one includes manufacturers who produced only prototype examples. When the British and European manufacturers are added to the North American total, the figure is likely to exceed 1200. Then one has to consider the Asian, African and South American manufacturers. Even without being able to arrive at a positive conclusion, the figure is obviously greater than one might at first think. But when consideration is given to the number of different *models* churned out by this unknown total of manufacturers — the final figure would certainly be beyond comprehension.

It is therefore a daunting task when a tractor historian is asked to identify the 'best' half-dozen *classic tractors* ever produced. What are the determining factors? How could one compare a 1905 tractor with a 1950 model? If pressed, one would possibly consider such icons as the Ferguson TEA20, the Caldwell Vale of 1910, the John Deere Model D, the Lanz Model DR of 1956, the Glasgow of 1919 and perhaps even the LA Case of the 1940s. A tractor that may not immediately spring to mind, but certainly worthy of consideration, is the Aultman Taylor 30-60.

The Aultman Taylor pedigree

The reason why the Aultman Taylor 30-60 could be considered in the short list of the 'best' tractors of all time is that its designers simply got it right! In 1911 this was a rarity. Designing and constructing a tractor was an imprecise science, as there were no precedents from which to learn. As a result tractors were unreliable (often hopelessly so) and usually grossly overrated. But not so with the Aultman Taylor 30-60. It was outstandingly reliable, with an *understated* horsepower availability.

In 1867 Cornelius Aultman and Henry Taylor registered the firm of Aultman Taylor & Co. located in Mansfield, Ohio. It was described as being a manufacturer of agricultural tools and machinery. The firm rapidly earned

a reputation for well engineered products and became a major producer of threshing machines and steam engines. Following a period of reorganisation the name was changed to the Aultman & Taylor Machinery Co. in 1892.

Around the turn of the century the company turned its attention to the development of a tractor powered by an internal combustion engine. After nearly a decade of experimentation the first Aultman Taylor 30-60 was trundled out of the factory in 1910.

It was obvious from the outset that the period of testing and development had paid off. The big machine performed its tasks without complications and attracted the attention of broadacre farmers throughout the expanding United States and Canadian grain belts. There was strong competition in these markets from Minneapolis, Rumely, Fairbanks Morse, International, Big Four and other prairie giants that emerged at that time. The common denominator shared by the designers of these tractors was that to, work in the wheat fields, tractors *had* to be big. Their design philosophies were thoroughly entwined with the experiences gained from their involvement with the great steamers of the previous century. Interestingly, the overwhelming consensus among veteran threshermen and tractormen who can recollect the days when these heavyweight internal combustion-powered tractors were still in use, is that the Aultman Taylor 30-60 was the pick of all the big machines.

A 1916 Aultman Taylor 30-60, owned by Dan Ehlerding of Jamestown, Ohio. Around 4000 of these big tractors were produced between 1910 until December 1923. Note the wooden handle for compressing air into the no.1. cylinder — see text. (Photo: I.M.J.)

The gentle heavyweight

Initially the Aultman Taylor 30-60 was equipped with a square-shaped induced draught radiator. This was abandoned in 1914 and replaced by a distinctive fan-cooled round tubular radiator with a capacity of 120 gallons.

The 4-cylinder cross mount engine was of Aultman Taylor's own design. It had a 7 x 9 inch bore and stroke, and developed its 60 belt h.p. at 500 r.p.m. The massive cast iron chassis also supported the transmission, which provided a single forward speed of 2.2 m.p.h. plus 1 reverse. Alternative gearing of 1.5 or 3 m.p.h. could be especially ordered.

Originally, the 30-60 was equipped with air-assisted start. The air pump was located to the left of the driver's control deck and had to be manually pumped by means of a long wooden handle. The air was fed into the no. 1 cylinder, where it was compressed. (It was first necessary to turn the engine so that the piston in no. 1 cylinder was at the top dead centre of its stroke.) Each cylinder received a squirt of petrol from an oil can through primers located on the cylinder head. The ignition system was switched on, the compressed gaseous mixture in no. 1 cylinder detonated, and the engine started. It was soon discovered, however, that the repeated sledge hammer detonation in the no. 1 cylinder was detrimental to the wellbeing of the piston, conrod and indeed the crankshaft. As a result, owners were encouraged to convert to an impulse magneto and abandon the air start. It was then necessary to crank the engine by means of a long lever inserted into a fitting at the flywheel end of the crankshaft. This was not the backbreaking task one might imagine, as the lever provided good leverage and the engine usually fired upon the first crank of the flywheel, providing the cylinder had first been primed with petrol. All subsequent versions of the 30-60 were equipped with impulse magnetos.

The belt driven off the flywheel provided the drive to the fan and water pump. Dan Ehlerding (pictured) recounts the time when a 30-60 shed the belt and it became entangled in the gears. The single forward gear engaged and the 12-ton tractor took off straight for the threshing mill which it had been powering. The team of threshermen had only seconds to leap from the mill before it was totally pulverised by the runaway tractor. (Photo: I.M.J.)

Aultman Taylor 30-60 tractors were normally supplied with an auxiliary petrol tank located on top of the left hand mudguard. This contained the petrol required for starting the engine. A 60-gallon kerosene tank was located underneath the operator's platform. Consumption of the 25 000 lb. tractor varied according to the load, but on ploughing applications around 6-8 gallons per hour was not uncommon.

In addition to the main engine clutch, a second clutch engaged the drive of the belt pulley. This enabled the pulley rotation to be engaged smoothly under load.

The massive 7 ft. 6 inch diameter rear driving wheels were supported on a 4.25 inch diameter axle. The standard wheel width was 24 inches but this was usually increased to 3 feet by fitting factory-provided extensions.

The lugging power of the 30-60 was legendary. The company's sales brochures claimed a drawbar pull of 8000 lbs, but an Aultman Taylor 30-60 submitted for testing at the Nebraska Tractor Test Laboratory on 30 June 1920 was found to have a sustainable drawbar pull of 9160 lbs at 2.38 m.p.h. (It is interesting to ponder that in 1955 an Oliver Super 99 diesel 6-cylinder tractor, one of the most powerful wheeled tractors of that year, returned a drawbar pull of 9212 lbs at 2.05 m.p.h.)

Country road engineers in the U.S.A. discovered that the Aultman Taylor 30-60 had sufficient power to easily pull *two* highway road graders at the one time. This represented a considerable saving in fuel and time as *both sides of the road* could be graded in one pass.

The big 30-60 was too large and too expensive for many farmers. Therefore Aultman Taylor also produced the 25-50, 22-45, 18-36 and the smallest of the range, the 15-30. But unquestionably the name of Aultman Taylor is usually fondly associated with the big 30-60.

In January 1924 the Aultman Taylor Machinery Company was purchased outright by the Advance Rumely Thresher Company of LaPorte, Indiana. This acquisition enabled Advance Rumely to erase its main competitor by discontinuing the production of Aultman Taylor tractors.

The Aultman Taylor 30-60 had a drawbar pull of 9160 lbs and could easily handle two of these 6-furrow mouldboard ploughs. Note the depth control levers for each of the mouldboards. It was common practice for a ploughman to ride each plough, amidst all the accompanying dust, and raise each mouldboard out of the ground at the headland. (Photo: I.M.J.)

CHAPTER 17

The Eye of the Beholder

I am obliged to Michael Garwood for his permission to reproduce these copyright prints in *The World of Classic Tractors*. This is a unique privilege, exclusive to this publication. The comments accompanying each of the works are purely my own interpretations. Other beholders may well interpret them differently.

Michael Garwood

The perception of classic tractors is an individualistic discernment that is frequently perplexing to define. A fondness for an aged Fordson might relate

Michael Garwood in his studio.
(Photo by kind permission of M. Garwood)

to warm memories of a grandfather. A regard for a Rumely could have its origins in nostalgic boyhood days; a love affair with a Lanz perhaps engendered because of its idiosyncratic single cylinder design.

Michael Garwood's perception of classic tractors relates to none of the these. His affection, especially for the abandoned derelict remnants of tractors, was inspired because they represent to him a phenomenon of shape, colour, light and shade, beauty and power, and majestic aloofness.

Michael Garwood is an artist of considerable ability, whose paintings are hung in the world's galleries extending from the National War Museum in Canberra to private and public galleries in London. His works are also represented in the collections of the Royal Australian Navy, Western Australia House (London), British Petroleum, the Shell Company, and by many other prestigious collectors of fine art.

His discovery of classic tractors occurred during the 1980s in Western Australia whilst on an extended exploration of the State's South West wheat belt. His objective was to portray on canvas old dilapidated buildings of national heritage importance. This is the type of work for which he is internationally famous.

One day in the dry, shimmering heat of summer Michael Garwood stumbled across a forlorn derelict tractor (identified later as a Turner Yeoman of England). It was parked alongside the scant remains of a tankstand, abandoned and forgotten. Extending from the elevated, rotting platform of the tankstand, like an accusing finger, was a nailed-on fence dropper to which were still attached two insulator cones — evidence of a long gone party telephone line. On a post below the insulators was a crudely fashioned wooden cabinet with its hinged door gaping ajar revealing its emptiness. Only an enlightened bushy would understand that it once contained a hand-crank telephone and its dry cell batteries.

The tractor was surprisingly intact, but the state of its tyres and electrical wiring was a clear indication that it had slumbered for many seasons at this once significant but now lonely and forgotten outpost. Someone had leaned the tail of a windmill against the tractor fuel tank. But who ... and when? Brackets welded to the the side frame revealed that a dozer blade had once been fitted. The paintwork had all but faded into oblivion, only a suggestion of the original green pigment remaining for detection by a perceptive eye.

Michael Garwood saw all these things and mentally enfolded them within the powerful ambience of mystery and intrigue. Here was a compelling new challenge for the artist, which later evolved into a passion. For a while the heritage buildings took second place to the search for more of these magnificent *Iron Bushrangers* as he referred to them. Long but gratifying months were spent in his studio translating the precise details of the Iron Bushrangers from his sketches and mind's eye onto canvas. The end results, now for all to see, are stunning. With his astute use of acrylic, Michael Garwood has captured the realism of the rust and decay of the tractors in a manner that few, if any, could replicate with a lens and sensitised film.

Although Michael Garwood may have only a passing knowledge of the history of classic tractors his contribution, through his magnificent paintings, to the rapidly growing interest in these artifacts is incalculable.

TWIN CITY 20-35 4-cylinder petrol engine, circa 1920.

The old tractor has outsurvived the gnarled desert oak leaning precariously in the background. Indeed, the Twin City appears in better shape than the neglected barbed wire fence. True, the rear mudguard is distorted — probably due to a lifetime of tight turns forcing the plough lift lever to jam against it — and the belt pulley has been purloined. Apart from that it seems to be fairly intact and straight. Whatever reason could there have been for its abandonment?

FORDSON MODEL N. 20 h.p. 4-cylinder, circa 1944.

A good guess is that the milk churns held the petrol for the tractor and the stationary engine, the 44-gallon drum served as the cooling radiator for the engine and the 4-gallon bucket was used for transporting water from the square tank behind the tractor. The rear lower sections of the Fordson mudguards have been roughly cut away, as they no doubt contributed to a build up of mud between the guards and the wheels. Some light-fingered person has appropriated the cast radiator top, but then someone else has kindly recovered the exhaust pipe and placed it securely against the seat. The rear tyres give the impression that they are still inflated, however this is only an illusion as they were obviously water ballasted and the water (not air) remains.

TURNER YEOMAN OF ENGLAND 40 h.p. V4 diesel engine, circa 1952.

One day in the dry shimmering heat of summer, Michael Garwood stumbled across a forlorn derelict tractor (identified later as a Turner Yeoman of England). It was parked alongside the scant remains of a tankstand, abandoned and forgotten. Extending from the elevated, rotting platform of the tankstand, like an accusing finger, was a nailed-on fence dropper to which were still attached two insulator cones — evidence of a long gone party telephone line. On a post below the insulators was a crudely fashioned wooden cabinet with its hinged door gaping ajar revealing its emptiness. Only an enlightened bushy would understand that it once contained a hand-crank telephone and its dry cell batteries.

The tractor was surprisingly intact, but the state of its tyres and electrical wiring was a clear indication that it had slumbered for many seasons at this once significant but now lonely forgotten outpost. Someone had leaned the tail of a windmill against the tractor fuel tank. But who - and when ? Brackets welded to the the side frame revealed that a dozer blade had once been fitted. The paintwork had all but faded into oblivion, only a suggestion of the original green pigment remaining for detection by a perceptive eye.

VICKERS AUSSIE 30 h.p. 4-cylinder petrol engine, circa 1925.

The remnants of one of the first Vickers tractors to come to Australia. (The term Aussie was dropped in 1927.) It is painfully obvious why this rig was abandoned. The absent bonnet and side panels, plus the cut down 44-gallon drum under the sump, all clearly indicate an engine failure and a forsaken attempt to carry out repairs. The presence and pattern of the non-original pneumatic tyres suggest the calamity occurred in the late 1930s. Whilst the reason for the proximity of the McKay stripper is self evident, the ancient wool press seems singularly out of place.

McDONALD IMPERIAL SUPER DIESEL TWB 1-cylinder 2-stroke engine, circa 1938.

The big single-cylinder semi diesel machine, with its dented rear mudguard and missing spark arrestor, is a sad sight for contemplation. The Sunshine grain bin is probably all that was worth salvaging from the stripper. The truck (is it a Chev or a Dodge?) with the remains of a charcoal burner hanging grimly to its side adds to the gloom. But the ultimate symbol of death is the butcher's gallows waiting patiently to hoist another carcase for final dissection.

WALLIS 20-30, circa 1929 and FORDSON MODEL F, circa 1927.

The two veteran tractors, still managing to appear dignified despite the scars inflicted from years of land battles, lead a parade of proudly erect, weary fence posts through a landscape of emptiness. Where are those now who owe them such a debt of gratitude?

HART PARR 18-36, circa 1926 and K.L. BULLDOG, circa 1950.

The hinges for the gate are there, but where is the gate? Probably borrowed by the same character who borrowed the front wheels from the Bulldog. The timber sun canopy with its shady hessian cover is also gone, as is the outer section of the chimney. The Hart Parr has fared better — at least it still has all its wheels — although its exhaust pipe has completely disappeared. The scattered drums remain, but are long past accommodating their respective fuels — the tarry crude oil for the Bulldog plus the petrol and kerosene for the Hart Parr. The inevitable cut-down 44 is also present. And what are these guides welded over the Bulldog's front axle? The young tree might have had all the answers, except it's already dead.

130 THE WORLD OF CLASSIC TRACTORS

FORDSON E27N MAJOR P6, circa 1952.

No ordinary Major this! It gives the impression from its regal, aloof bearing that it is haughtily conscious of its optional extra Perkins 6-cylinder diesel engine, plus lighting kit — including special rear work lamp! The only items missing are the headlamp bulbs and lenses. Some considerate person has tied a 'roo skin around the exhaust pipe to keep out the rain. The tyres could still be O.K. Even the water temperature gauge is still in place and can be seen peeping from the rear of the radiator header tank. By the look of the fuel drums, though, it certainly wasn't parked there yesterday.

CHAPTER 18

'Home Grown' Australians

The U.S.A. had been colonised for 200 years before Australia's development really got underway. Right up to the 20th century the fledgling colony, situated at the bottom of the world, depended almost entirely upon Mother England to provide everything from rabbit traps to steam engines. It took six months to receive a reply from Britain to a letter posted in Sydney, and usually twelve months to receive machinery or spare parts from the time the order was posted off. By necessity, therefore, Australians became *great innovators*. This was particularly true in the bush, where isolated pastoralists and farmers had to depend upon their own initiative to repair broken machinery and often had no option but to invent home-made implements or gadgets. This characteristic of innovation and inventiveness was manifested in some of the unique Australian tractors that were to emerge as the 20th century unfolded.

Caldwell Vale

Manufactured at Auburn, N.S.W. by the Caldwell Vale Motor & Tractor Construction Co. Ltd., the 4-wheel-drive heavyweight was considered by

1910 Caldwell Vale. (Photo from an original glass negative, circa 1910, courtesy G. Crooks)

many to be the world's most advanced tractor in 1910. It featured the firm's own design 4-cylinder 680 cu. inch petrol engine, having a 6.5 inch bore and stroke and developing 80 h.p. at 600 r.p.m. The power was delivered to the four driving wheels through an enclosed 3-speed gearbox. The tractor also featured 4-wheel steer and 4-wheel brakes. *There was simply no other tractor produced in the world in 1910 that could equate the advanced design of the Caldwell Vale.*

Regrettably, the company folded in 1913 as the result of a court judgement found against it involving a substantial claim for damages and legal costs. It is interesting to speculate what might have developed had Caldwell Vale managed to stay afloat. Tractor historians and enthusiasts outside Australia largely remain in complete ignorance of this historically important tractor.

It is believed that no Caldwell Vale tractors survive intact today. Fortunately, however, there are remnants in both South Australia and Queensland that are likely to undergo eventual restoration.

McDonald No. 148

A.H. McDonald of Richmond, Victoria, pioneered tractor production in Australia when the firm released its 20 h.p. Model EA in 1908. The unit pictured — No. 148 — is believed to be a Model EB. The photograph was taken in the Rockhampton district of Queensland in 1914. In K.N. McDonald's superb book *A.H. McDonald — Industrial Pioneer* there is a photo on page 80 of tractor No. 147. The only difference between the two tractors seems to be that No. 147 has Pedrail tracks fitted to the rear wheels and was designated the EC.

1914 McDonald No. 148. Note the dreaded prickly pear growing in the foreground. (Photo: part of a historic collection housed in the John Oxley Library, Brisbane)

It appears that most of the McDonald EB tractors were fitted with tower radiators, whereas both Nos. 147 and 148 had the more efficient fan-cooled tube type radiator. This was undoubtedly to obtain more effective cooling in tropical conditions. The engine in No. 148 was the McDonald 2-cylinder type DB rated as being 25 h.p.

Jelbart No. 123

Jelbart Bros. of Ballarat, Victoria, were arch rivals to the A.H. McDonald Co., Richmond, as both firms produced stationary and portable engines, tractors and road rollers. Jelbart entered the tractor business six years behind McDonald, in 1914, and continued production until 1926. During that time a range of technically interesting Jelbart engines was installed in the tractors.

In 1909 Jelbart patented a single-cylinder 2-stroke low-compression engine which *scavenged* its intake air from the crankcase. Unlike later 2-stroke tractor engines with scavenging systems (e.g. Lanz, Marshall and Avance) the Jelbart featured a *stepped piston.* Its lower end (skirt) had a larger diameter than the main piston body. The reason for this was to create greater suction when drawing air into the crankcase, followed by quicker purging when it was expelled into the combustion chamber.

Jelbart No. 123, circa 1917, photographed during its restoration in 1995 at the Red Hill Museum, Balmoral, Victoria. The mechanical work had been completed and the tractor ran magnificently (for a Jelbart). (Photo: I.M.J.)

The original single-cylinder Jelbart tractor engine featured a 7 inch stroke with a 7 inch diameter bore widened to 9 inches at the lower end of the cylinder, in order to accommodate the stepped piston. It was claimed (probably conservatively) to develop 14 h.p. at around 400 r.p.m.

The Werribee Tractor Trials

Conducted by the Royal Agricultural Society
and the University of Melbourne

At the Experimental Farm on Wednesday and Thursday, September 18 and 19, 1918.

After the most exhaustive test the results were recorded and published in the Press.

The following extract appears in the "Age" of Thursday, 26th September, 1918 :—

Competitor	Acres Ploughed	Time Hours	Fuel Gals.	Water Gals.	Acres Hour	Fuel Acre
Bates Steel Mule) Clutterbuck Bros.)	3.60	3.67	9.30	8.07	0.95	2.66
Imperial E.A.A.) McDonald & Coy.)	3.64	3.78	10.94	4.62	0.94	3.10
Waterloo Boy) Mitchell & Coy.)	1.40	4.69	7.68	0.56	0.30	6.41
Sunshine A.) H.V.McKay)	1.79	4.69	6.40	9.69	0.38	3.02
Imperial Light Weight) McDonald & Coy.)	2.12	3.11	6.27	5.25	0.68	2.49
Mougul International) Harvester Coy.)	2.82	4.50	12.06	10.61	0.63	4.28
Crude Oil Tractor) JELBART'S PTY.LTD.)	3.57	2.82	6.26	6.85	1.27	1.75
Sunshine O) H.V.Mc.Kay)	3.63	4.45	7.73	15.60	0.82	2.13

The fuel used in all cases was "Borneo Cross" Kerosene.

The "Jelbart" was the only Tractor which was prepared to conduct the test using Crude Oil Fuel, and the makers are prepared to prove at any time that they can reduce the costs of the fuel consumed as set out above by 40 to 60 per cent.

The figures above set out absolutely prove that the "JELBART" is the most efficient and economical Tractor on the Australian market.

This advertisement appeared in several rural newspapers in October 1918. Note the incorrect spelling of the International MOGUL.

Included among the different Jelbart engines fitted to the tractors was a 45 h.p. model which had a 12 inch bore but was also fitted with a horizontally opposed cylinder with an 8 inch bore. The small piston served to balance the engine and increased the pressure in the crankcase, thus supercharging the scavenging system. (Note: This principal was re-invented by David Brown in 1956 and incorporated in the 2D engine. Also, Electrix Ltd. of Dagenham, England released a tractor at the Smithfield Show of 1949 powered by a 45 h.p. 2-cylinder 2-stroke diesel engine with four pistons, to aid scavenging.)

Jelbart tractors were provided with a leather segmented belt to deliver the engine power to the 3-forward and single-reverse-speed transmission. The belt was tensioned by a jockey pulley which acted as the clutch. To release the 'clutch' the operator moved a lever which relieved the belt tension. The system was crude but efficient.

Characteristic of every Jelbart tractor was the racket emitted each time the cylinder fired. If a modern environmental officer approached a Jelbart with a decibel testing instrument, both he and his instrument would be rendered inoperable due to the shock of the noise.

A few Jelbart tractors were imported into New Zealand. A restored example is in the hands of a collector at Taurangu. Two others — an original and a wreck — are at Leven, and there are two in Southland. Apart from New Zealand, the only other country to receive a Jelbart tractor was Kenya, where in 1926 one of the final units made was purchased by a local Njoro plantation owner.

McDonald Imperial TWB

A.H. McDonald & Co., in great secrecy, acquired a Lanz Bulldog tractor in 1928 for the sole purpose of stripping it down so that its engine components could be measured and sketched by the McDonald design engineers. Following its embarrassing and costly association with the Swedish Avance tractor (see *Agricultural Tractors — Their Evolution in Australia* by I.M.J.) the firm was anxious to push ahead, with a degree of urgency, the drawings for a new tractor. It was considered that the engine ideally should be similar to the single-cylinder 2-stroke hot-bulb principal as fitted to the highly successful Lanz Bulldog. (By an amazing coincidence, at exactly the same time as the McDonald team was surreptitiously studying the Lanz, a similar undertaking was taking place at the Britannia Iron Works in Lincolnshire by William Marshall Sons & Co. As a result of its clandestine studies, the single-cylinder Marshall 15-30 was released in 1930.)

The single-cylinder McDonald T type engine was ready for installing in the first of the new tractors in 1930. It had been decided not to emulate the Lanz transmission, probably on account of the expensive tooling and casting costs which would have been necessary. Instead, an arrangement had been made with the Advance Rumely Thresher Co. of Indiana to purchase Rumely transmissions.

The first four of the new tractors were designated the Model TX, on account of the T engine and X type imported Rumely transmission. Future tractors were equipped with McDonald-made Rumely rear ends and were identified as the TW, followed by the TWB in October 1931. The drive from the engine to the transmission was by a roller chain, as were the final drives to the axles. As earlier stated, this was a less costly system to manufacture than if the Lanz transmission design had been adopted.

The engine was mounted with the hot-bulb cylinder head facing rearward towards the operator. (The Lanz and Marshall engines faced forward.) It was necessary to pre-heat the hot bulb with a blowlamp prior to starting. Being a valveless 2-stroke, when the flywheel was swung the engine could run in reverse. This necessitated cutting the flow of the crude oil fuel until the engine was on the point of stalling, then opening the fuel lever fully in the hope that the engine would bounce in the opposite direction. With practice, this could be achieved quite readily. The 9.25 x 10 inch bore and stroke produced 36 h.p. at 535 r.p.m.

As a special inducement, McDonalds offered Australian farmers an attractive deal whereby the tractors could be returned to the factory for a complete restoration and upgrade, rendering them equivalent to the latest model.

The McDonald single-cylinder tractors had immense character and are greatly sought after now by tractor collectors. The 1934 TWB pictured has been meticulously restored by Alex and Ron Grosser and is on display at the Gunnedah Rural Museum. (Photo: I.M.J.)

A view from the cockpit of Alex and Ron Grosser's TWB. Note the hot bulb, which had to be pre-heated with a blowlamp prior to starting. The Rumely transmission housing is clearly pictured, including the shaped cast top which covers the chain drive final drive sprockets.
(Photo: I.M.J.)

The old flour mill at the Temora Rural Museum is a fitting background for the 1929 Ronaldson Tippett Super Drive. The museum's tractor expert, Ron Maslin, is at the controls.
(Photo: I.M.J.)

Ronaldson Tippett Super Drive

The origins and details of the Ronaldson Tippett Super Drive tractor are discussed in Chapter 30. However, it is interesting to note that the Ballarat firm produced its first prototype tractor in 1910. This was the same year as the Caldwell Vale entered production, following prototype experimentation which commenced in 1907. The 20 h.p. twin-cylinder Ronaldson Tippett design was archaic compared to that of the Caldwell Vale. The comparison is not really relevant, though, as Ronaldson Tippett committed its energies to the production of stationary and portable engines. Also, the firm endured until 1972 — Caldwell Vale folded in 1913.

By 1924 Ronaldson Tippett had produced in excess of 5000 engines and was actively competing in some export markets. It is not surprising that, with its capacity fully utilised producing engines, the company decided to import the Illinois Super Drive to rebadge and sell to Australian farmers as a Ronaldson Tippett product. This turned out to be a wise decision as the Super Drive tractors proved reliable wheatland machines, having been thoroughly tested in the U.S.A. This saved Ronaldson Tippett time and money by not requiring the firm's involvement in prototype testing.

KL Bulldog

Yet another Australian manufacturer was sufficiently impressed by the simplicity of the German Lanz Bulldog tractor and its reputation for reliability that it decided to produce an Australian equivalent. Kelly & Lewis was given encouragement by the Federal Government with its promise of material availability — this in the post-war period of acute steel shortages. The nation could not import sufficient tractors in the mid-1940s to meet the demands of farmers being pressed to open up new country for the production of grain. Kelly & Lewis of Melbourne had an association with the German product prior to World War II and had a comprehensive inventory of Lanz parts in its store from which to copy.

The first KL Bulldog came off the production line of the specially constructed Springvale Works in March 1949. Only an expert could have identified the difference between the KL and the German original. The tractor was a virtual carbon copy of the D8506 Lanz. The single-cylinder 2-stroke semi diesel Australian Bulldog engine had an 8.86 x 10.24 inch bore and stroke and developed its 44.25 belt h.p. at 600 r.p.m.

Gary Skinner of Uralla, N.S.W. has painstakingly returned his 1949 KL Bulldog to original condition. (Photo: I.M.J.)

Unfortunately for Kelly & Lewis, in the early 1950s Lanz introduced a complete new range of single-cylinder Bulldogs fitted with a light alloy piston plus redesigned scavenging system and combustion chamber. The new imported tractors immediately rendered the KL old-fashioned. Optimistic figures for the KL were never attained and only a mere 900 were built. The firm ceased manufacturing the KL Bulldog in 1954 but was obliged to spend the next six years endeavouring to dispose of stocks.

Like so many small production run tractors — and in particular because it is an *Australian Bulldog* — the KL is today greatly treasured by both Australian and overseas collectors. The few European-based Bulldog collectors who are fortunate to have a KL consider it to be *the most collectable of all Bulldogs*. Thus the value of a KL in Europe is many times greater than in Australia.

Howard Kelpie

It is paradoxical that Clifford Howard, whose company produced the most mundane of all Australian tractors, gained a degree of international prominence unequalled by any other Australian tractor manufacturer.

The DH22, launched in 1930, was the first *conventional* tractor produced by Howard Auto Cultivators Ltd. of Northmead, N.S.W. It followed a comprehensive range of motorised cultivators stemming back to 1921. The DH22, along with other Howard implements, was marketed in numerous countries around the world including Britain and the U.S.A.

In 1938 Clifford Howard also set up a completely separate company in England registered as Rotary Hoe Cultivators Ltd. Subsequently he established Rotary Hoes Ltd. and the Platypus Tractor Co., both in England. The latter firm produced an interesting range of lightweight crawlers including a wide-track version designed exclusively for operating in swamp country.

It is not widely known that Rotary Hoes Ltd. acquired the celebrated and historic John Fowler & Co. (Leeds) Ltd. in May 1945. Clifford Howard became one of six directors and was appointed managing director. The following year the Thomas W. Ward Group, which had for a decade been associated with Fowler, extended an offer to purchase Fowler from Rotary Hoes Ltd. The deal, involving complex share transactions, was accepted and Wards took over Fowler in December 1946. (This acquisition effectively joined the Fowler and Marshall conglomerates.)

Whilst all the foregoing corporate intrigue was taking place in England, Howard Auto Cultivators at Northmead introduced a diminutive market garden tractor named the Kelpie. It was powered by a single-cylinder 4-stroke air-cooled Howard engine developing 5.2 h.p. at 3200 r.p.m.

1952 Howard Kelpie owned and neatly restored by John Mullington. (Photo: I.M.J., taken at 1996 Rusty Iron Rally, Wauchope, N.S.W.)

Few Kelpie enthusiasts realise that the little unit was a direct copy of a tractor made in Stratford, Connecticut by the Beaver Tractor Co. Inc. The engine in the American Beaver was a Wisconsin AKN. Howard manufactured this engine in Australia during World War II for driving lighting plants for the Australian Armed Forces. After the war Howard continued making the engine, but under the Howard brand name. Interestingly, the engine is still in production by Wisconsin in the U.S.A., where it is known as the BKN.

Owing to the simplicity of the Kelpie's design, those that remain today can generally be put into good working order at very little cost. Many still perform regularly on hobby farms and commercial vegetable gardens throughout Australia.

A 1955 advertisement for the Platypus 30 tractor which appeared in a number of British farming publications. Howard distributed the Platypus through his associated company, Rotary Hoes Ltd.

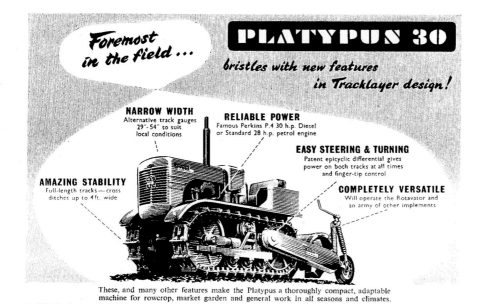

Dinkum Digger Mark II

At first glance the Dinkum Digger might appear to have no place in a chapter entitled 'Home Grown Australians' but the Dinkum Digger Mark II is very much Australian — although with a Scottish ancestry.

Despite all the extravagant claims made by several manufacturers of hydraulic-powered digging machines to have pioneered the original concept, in fact the design of the very first *fully hydraulic* backhoe must be attributed to a Scot named Robin Ewen. His prototype was built and tested at Cupar, Fife in 1952. Following its successful trials, Ewen entered into an agreement with brothers Carlton and Roy Whitlock of Whitlock Bros. Ltd., Great Yeldham, Essex to manufacture and distribute the Dinkum Digger, principally for mounting on Ferguson TE and Fordson Major tractors. (Note: The word 'dinkum' is generally considered by Australians to be of Australian origin, but it is in fact an old Scottish dialect word.)

E.P. Lough & Co. of North Sydney negotiated the Australian agency rights for the Dinkum Digger with Whitlock Bros. The majority imported into Australia were retailed by British Farm Equipment Pty. Ltd. who were the distributors for Ferguson tractors (and subsequently Fiat) in Victoria and New South Wales. When Massey Harris Ferguson took responsibility for the Ferguson distribution in Australia, E.P. Lough & Co. created its own retail subsidiary in 1959 — Lough Equipment Pty. Ltd. at Artarmon, N.S.W.

Although the larger Whitlock Major and 60 backhoes were coming on stream by that time, it was felt there was still a market for Dinkum Diggers in Australia. In 1961, with approval from Great Yeldham, Mr. Eric Lough of E.P. Lough & Co. arranged for the engineering firm of Walters & Co., Auburn, N.S.W., to manufacture an improved Australian version of the Dinkum Digger. The new machine featured individual hydraulically

A Dinkum Digger Mark II purchased by H. Walshaw, a Sydney Western Suburbs plant hire operator. It was mounted upon a Massey Ferguson 35 tractor fitted with a Horndraulic front end loader.
(Photo: I.M.J., circa 1961)

controlled stabilisers, which replaced the original dropdown sprags. The digging depth was increased from 7 to 9 feet and the steel rope slew mechanism improved.

The Dinkum Digger Mark II, as it was termed, was sold mounted on a Massey Ferguson 35 tractor as a complete unit and usually with a front end loader. Some were retailed by the new industrial Massey Ferguson dealer, Cumberland Tractors Pty. Ltd., Auburn, N.S.W., but the majority were purchased by plumbers and drainers through Lough Equipment Pty.Ltd. on tractors acquired on a wholesale basis by Lough from Cumberland Tractors.

The Australian production of Whitlock Dinkum Diggers was discontinued in 1964 when Lough Equipment Pty. Ltd. was appointed Australia's first distributor of J.C.B. loader backhoes.

Chamberlain Super 90

In 1962 there were some splendid heavyweight wheeled tractors available to broadacre farmers around the world. These included the Case 930, Oliver 1900D, Fiat 88R, John Deere 4010 and the Minneapolis Moline GB1 — but only Australian farmers had the opportunity of experiencing the outstanding Chamberlain Super 90.

Chamberlain Industries Ltd. of Welshpool, Western Australia introduced its first tractor in 1949. This was the unique 40K powered by Chamberlain's own twin-cylinder, transversally mounted, horizontally opposed 40 h.p. engine. The Chamberlain 40K and the majority of ensuing models were designed specifically for the Australian grain farmer.

Protected by a tariff bounty system, Chamberlain tractors were priced extremely competitively against imported machines. But as the American manufacturers introduced increased sophistication and larger engine capacities into their tractors, Chamberlain was obliged to keep pace. As a consequence the Super 90 was released in 1962 and was (and remained) the best tractor ever to come out of the Welshpool plant. Although lacking some of the refinements of the big American machines (it still had a hand-operated clutch) its field performance was exceptional.

A 1964 Chamberlain Super 90 Mark II pulling a 22-disc International Model A-1 plough. No matter how hard the ground, the Super 90 handled the big plough with ease at around 5.5 m.p.h. In light country the tractor had the capability of pulling two such ploughs hooked in tandem. (Photo: I.M.J., 1978, at 'Chelmsford', Rowena, N.S.W; driver Peter Crofts)

Note: The drawbar pull of the Super 90 was 8406 lbs at 3.1 m.p.h. The Oliver 1900D, the most powerful of the other tractors previously mentioned, under test at Nebraska returned a 12 475 lbs pull, but at the *much slower speed* of 1.38 m.p.h. and *with 4740 lbs of additional ballast* in the form of stacked weights clamped to the rear axles. At 4.95 m.p.h. the pull of the Oliver was reduced to 6282 lbs The figures can be confusing, but the important fact here is that the Super 90, which in *normal operating trim* weighed around 1500 lbs more than the Oliver, had the ability to transfer its horsepower onto the ground! It is fascinating to contemplate that the Oliver 1900D was powered by a GM 4-cylinder 2-stroke supercharged engine of *212 cu. inch capacity* and the Chamberlain Super 90 had the GM 3-cylinder 2-stroke supercharged engine of *213 cu. inch capacity.*

It should also be noted that the Super 90 Mark II was given an additional engine horsepower boost from the original 85 to 100, which rendered the tractor even more aggressive in the drawbar pull department.

The Super 90 had nine well spaced gears and (as previously stated) a hand-controlled over centre clutch which was a joy to use. The offset upholstered bucket seat provided a relaxed driving position and the right hand mudguard was perfectly positioned to act as an armrest. The front axle had a soft leaf spring suspension, adding to the operator's comfort.

The Chamberlain Super 90 was phased out on economic grounds. The cost of its massively rugged construction and the relatively high-priced imported GM 3-71 engine proved uneconomical in relation to a realistic retail price. The replacement in 1968 was the Countryman 354 powered by a Perkins 6-354 engine. Although more modern in appearance and in its own way a good solid tractor, the Countryman was no substitute for the Super 90. The Super 90 *was a classic in its own time.*

The author standing alongside his Chamberlain Super 90, upon which he recorded many hundreds of hours in the 1970s, broadacre farming at 'Chelmsford', Rowena, N.S.W. (Photo: M. Daw)

CHAPTER 19

Ferguson — What *Really* Happened!

Classic tractor enthusiasts are sometimes perplexed by the relationship of different tractors bearing the name FERGUSON. This chapter aims to untangle the confusion.

The David Brown episode

In 1936 two prominent industrialists commenced a business association which was to have far-reaching effects upon the future of the farm tractor.

Harry Ferguson, pioneer aviator, champion racing driver and brilliant farm machinery philosopher, had designed a revolutionary tractor incorporating his patented Ferguson System of 3-point linkage implement attachment and draught control. David Brown, 'gentleman farmer' and chairman of one of Britain's major engineering firms, agreed to establish a production plant in which he would manufacture Ferguson's tractors. The tractors were to be known as the Ferguson Model A (see 'David Brown the Innovator', Chapter 12).

The initial 500 units, released in 1936, were powered by a side valve Coventry Climax engine but later machines were equipped with a 2010 c.c. David Brown petrol engine which developed 20 h.p. at 1400 r.p.m.

The Model A proved convincingly that a lightweight tractor incorporating the Ferguson System of weight transfer and draught control could handle a 2-furrow plough that normally would have required a tractor of double its

LEFT 1936 Ferguson Model A manufactured by David Brown Tractors Ltd. (Photo: I.M.J., courtesy of the Ulster Transport Museum, Holywood, Northern Ireland)

RIGHT Rear view of Ferguson Model A showing a 3-point linkage mounted ridger. (Photo: I.M.J., courtesy of the Ulster Transport Museum, Holywood, Northern Ireland)

weight. There was a catch, though! It could only achieve this when matched to the specially designed Ferguson 3-point linkage implements.

British farmers in the 1930s found the going tough. The effects of the Depression years tended to remain longer in rural communities. As a result, few farmers were in the financial position of being able to afford a new tractor. Those who managed to convince their reluctant bank manager to approve a tractor loan could rarely summon the necessary courage to also request additional funding for the purchase of a new set of implements. The whole point in investing in a Ferguson was to obtain the benefits of the *Ferguson System* which (as previously stated) could only be obtained by also owning the matched implements.

Accordingly, sales of the Model A fell well below expectations. Profits from the venture were not forthcoming. It was then considered that having David Brown Tractors Ltd. as the manufacturer and Harry Ferguson Ltd. as the design and marketing operation was doubling up on many of the administration costs. So the two operations amalgamated into Ferguson Brown Ltd. But even this could not turn the losses into profits.

By 1938 the two great men were squabbling. Brown insisted that to obtain volume sales a higher horsepower tractor was required with four forward gears, instead of the three as in the Model A. Brown also believed a power-take-off shaft should be fitted as standard equipment. Ferguson argued strongly against any such changes.

In 1939 the partnership was dissolved, accompanied by heated words from Ferguson. A total of 1350 Ferguson Model A tractors had been produced — well below the anticipated figure.

Brown, who had been working independently upon his idea of a more powerful tractor, unveiled the all-new and technically advanced David Brown VAK 1 at the British Royal Agricultural Show in 1939, just weeks before the outbreak of World War II. Ferguson, who had already shipped a Model A across the Atlantic to vehicle magnate Henry Ford, set sail for the U.S.A. where he was greeted warmly by his old friend.

The Henry Ford episode

Henry Ford was totally convinced by the clever Ferguson System design demonstrated by the astute Harry Ferguson. He required little persuasion to issue the necessary instructions to his design team to incorporate the patented system into his proposed new Ford 9N tractor. He even acceded to Ferguson's wish to have the tractor painted in Ferguson grey. The production 9N tractor was badged a 'Ford' but, at Ferguson's insistence, bore an additional bonnet emblem which read 'Ferguson System'.

The Ford 9N was powered by a derivation of the engine installed in the military Ford Jeep. This was a 4-cylinder side valve petrol-powered unit of 120 cu.inch. displacement which produced 24 h.p. at 2000 r.p.m. In fact it was virtually half a Ford V8 and used many interchangeable components.

A 1939 Ford 9N on display at Temora Rural Museum. To the uninitiated the 9N is frequently mistaken for a Ferguson TEA 20. An easy way of determining the difference is to note the opening panel on top of the bonnet and the wheel design of the Ford. (Photo: I.M.J., with kind permission of Temora Rural Museum)

The Ford 9N was usually referred to as the 'Ford Ferguson'. Sales were considered excellent. The deal between the two entrepreneurs, known in legal circles as the Handshake Agreement, stated in essence that Ford would build the tractor but Ferguson would be responsible for its distribution through the marketing firm of Ferguson-Sherman Co., which included the brothers Eber and George Sherman. Ferguson also retained control of implement design.

In 1941 Ferguson had a disagreement with the Sherman brothers and terminated his arrangement with them. He promptly established yet another marketing organisation, which he alone controlled.

The little grey 'Ford Ferguson' continued to sell well and farmers in America and overseas, including Australia and New Zealand, appreciated the advantages of the Ferguson System. An economy wartime version was introduced in 1942 known as the 2N. Because of component shortages it was not fitted with a starter and generator. Nor was it mounted on pneumatic tyres but supplied with steel wheels, which at that time were losing their popularity.

The turbulent period

Henry Ford's grandson, Henry Ford II, took over control of the Ford tractor manufacturing division from his ageing grandfather in 1943. Harry Ferguson, true to form, had a disagreement with the grandson over a variety of matters.

At about the same time Harry Ferguson also quarrelled with Lord Perry, who was Ford's senior man at the English Dagenham plant. Lord Perry would not agree to Ferguson's insistence that the well advanced plans to launch the Fordson Major in 1945 be abandoned. Ferguson wanted the Dagenham works to produce the 'Ford Ferguson' in place of the Major.

Feeling somewhat frustrated, Ferguson approached Sir John Black of the Standard Motor Co., who agreed to utilise a vacant factory at Coventry, in the English Midlands, for the purpose of manufacturing Ferguson (as distinct from 'Ford Ferguson') tractors. Ferguson had already approached the British Government and had been granted the necessary licences to obtain steel for the project. Thus the Ferguson TE series was born in 1946, firstly with a Continental side valve engine then a Standard 4-cylinder overhead valve engine. The latter was being used with great success in Standard Vanguard and Triumph TR2 cars.

In the meantime, back across the Atlantic Ford had produced 306 221 9N/2N tractors equipped with the Ferguson System. Henry Ford II launched a new model in 1947, somewhat cryptically named the 8N, which retained the Ferguson System *without the approval of Harry Ferguson!* Ferguson erupted, and immediately took legal proceedings against Ford. The Ford defence was that the company was entitled to go on using the Ferguson System as this was in complete accord with the Handshake Agreement.

Thoroughly enraged, Ferguson in a remarkably short time established a factory in Detroit, where he manufactured the TO Ferguson and sold it in direct opposition in the United States to the new Ford 8N. Prior to the TO coming on stream he had rushed shipments of the TE from England in order to lose no time in commencing a vigorous marketing campaign against the Ford 8N.

The court case between Ferguson and Ford was at that time the largest civil suit ever to come before a court. Ferguson sued Ford for $340 million. The case dragged on for years and was finally settled out of court in Ferguson's favour. He received $9.25 million, of which nearly half went towards legal expenses.

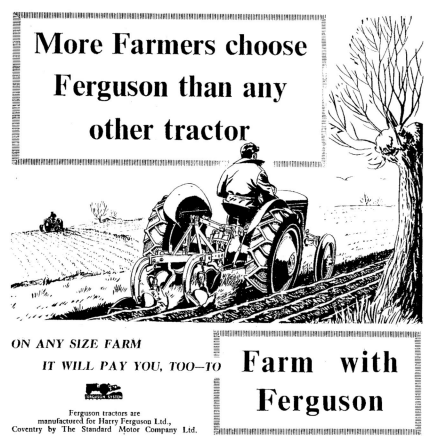

Ferguson advertised extensively in farming periodicals. This example appeared in English-speaking countries in 1952.

The rest is history. David Brown achieved huge success with his post-war Cropmasters and their derivatives. The Ferguson TEA series became the most successful tractor of the post-war era. The Dagenham-built Fordson E27N Major, despite its outdated concept, for a while was Britain's top selling and top export tractor. The U.S.-built Ford tractors became one of that country's market leaders.

Today, throughout the world, every tractor equipped with 3-point linkage and draught control owes at least an acknowledgement to the memory of the great Harry Ferguson and this turbulent period.

The Ferguson System logo was cleverly designed and instantly recognisable by farmers in over 80 countries around the world.

'Ferguson System' Equipped Tractors — 1936-1964

Type A	David Brown built 1936-39
9N	Ford built 1939-47
9NAN	Ford built 1939-47 with V.O. engine for U.K.
2N	Ford built 1942-45 utility version
TE20	Standard built 1946-48 with Continental Z120 engine
TO20	Ferguson Detroit built 1948-51 with Continental Z120 engine
TEA20	Standard built 1947-56 with Standard petrol engine
TEB20	Standard built 1946-48 with Continental Z120 engine, narrow
TEC20	Standard built 1948-56 with Standard petrol engine, narrow
TED20	Standard built 1949-56 with Standard V.O. engine
TEE20	Standard built 1949-56 with Standard V.O. engine, narrow
TEF20	Standard built 1951-56 with Standard diesel engine
TO30	Ferguson Detroit built 1951-54 with Continental Z129 engine
TEH20	Standard built 1950-56 with Standard Zero Octane engine
TEJ20	Standard built 1950-56 with Standard Zero Octane engine, narrow
TEK20	Standard built 1952-56 with Standard petrol engine, Vineyard
TEL20	Standard built 1952-56 with Standard V.O. engine, Vineyard
TEM20	Standard built 1952-56 with Standard Zero Octane engine, Vineyard
TEP20	Standard built 1952-56 with Standard petrol engine, Industrial
TER20	Standard built 1952-56 with Standard lamp oil engine, Industrial
TET20	Standard built 1952-56 with Standard diesel engine, Industrial
TO35	Ferguson Detroit built 1954-57 with Continental Z135 engine, Rowcrop
FE35	Standard built 1956-58 with Standard petrol engine
MF35	Standard/Massey Ferguson built 1958-59 with Standard petrol engine
MF35	Massey Ferguson built 1959-62 with Perkins 3.152 diesel engine
MF35X	Massey Ferguson built 1962-64 with Perkins A3 152 diesel engine

Note: TE and TEA tractors were built side by side up to serial number 48000.
12 volt electrical system was introduced from serial number 250001.
85 m.m. engine was introduced in 1951 from serial number 172501.

CHAPTER 20

The Galloway Farmobile

The following is a true story recounted to the author by noted farmer and classic tractor collector Kenny Kass who comes from Dunkerton, Iowa.

The disagreement

John and Jack Smith (not their real names) were two brothers who in 1916 farmed 200 acres of corn out on the flat plains of Iowa. John was fascinated with the technology and potential of the increasing number of clattering tractors that were disturbing the tranquillity of the normally peaceful countryside. Within a couple of days' buggy ride from their farm were a number of engineering shops producing these newfangled contraptions. There was Heider at Carroll, the Waterloo Gas Engine Company at Waterloo, plus the Rock Island Plow Co. across the Mississippi in Illinois. But the tractor that captured the imagination of John Smith was the Farmobile built by the Wm. Galloway Co. of Waterloo.

Wm. Galloway had become disenchanted with his job as a farm implement salesman. In 1902 he established a business in Waterloo, initially selling

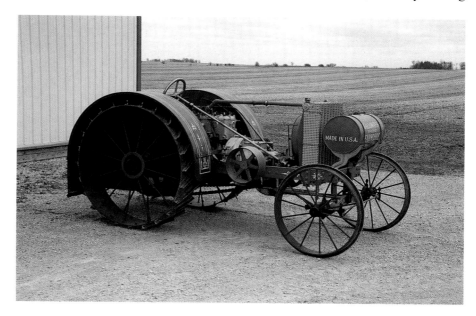

A view of the Galloway Farmobile owned by Kenny Kass, Iowa.
(Photo: I.M.J.)

Cadillac cars, but later dabbling with a car of his own manufacture. Following a number of financially tumultuous years he designed and introduced the Farmobile tractor in 1916.

The Galloway Farmobile was powered by an engine manufactured by the nearby Dart Truck & Tractor Corp. It was a 4-cylinder petrol-powered side valve unit with a 4.5 x 5 inch bore and stroke and developed 20 brake h.p. The tractor was equipped with 1 forward and 1 reverse gear and weighed around 2.5 tons.

This was the tractor favoured by John Smith. John had a problem, however. His brother Jack steadfastly refused to entertain the idea of introducing one of these 'monstrosities' onto the farm. For some months the brothers rarely spoke to each other as they simmered with determination. Finally Jack yielded to his brother, as John had threatened to sell up his share of the farm if a tractor was not purchased.

Enter the Farmobile

In due course the gleaming new red and green tractor arrived at the railhead and was driven to the Smith farm. It only worked a few short weeks when the arrival of the harsh Iowa winter signalled a halt to all outside farm activities until the spring.

The left hand view of the 1916 Farmobile clearly shows the Dart engine, serial no 519, and the transmission and clutch mounted between the engine and the radiator. Kenny Kass is at the wheel. (Photo: I.M.J.)

The tractor was put in the shed and its water drained.

During that winter John Smith contracted an illness and died. When spring came, Jack harnessed up his horses and left the tractor in the shed. There it remained for a quarter of a century under lock and key. Jack Smith also ultimately ascended to the big paddock in the sky and the farm, together with the tractor, was sold. An antique vehicle collector, appreciating the uniqueness of the Galloway Farmobile, purchased it — not to use, but to retain as an interesting antique. This occurred around 1950.

Years later a young Kenny Kass indicated to its new owner his desire to one day own the Farmobile. The years rolled by and Kenny Kass exhibited more than a passing interest in classic tractors. Quietly and progressively he was putting together an outstanding private collection of rare machines. Eventually the Galloway Farmobile passed into his caring hands.

Extraordinarily, the only work ever performed by the tractor was back in 1916 when it was driven for a few weeks by John Smith. As a consequence this octogenarian artifact is in *as-new condition*.

It is historically significant that the Wm. Galloway Co. also made a smaller tractor which was sold in Britain under a licensing arrangement with Henry Garner Ltd. of Birmingham (see Chapter 28). Apparently Galloway railed 200 of these tractors to New York for the sea voyage to England. They were never shipped, owing to the ocean freight priorities of World War I. This had a serious financial implication for the Wm. Galloway Co. from which it never fully recovered. The company went out of business in 1920.

The rear view of the Galloway Farmobile. Note the location of the crank handle. (Photo: I.M.J.)

CHAPTER 21

As British as Winston Churchill

The name William Marshall, Sons & Co. of Gainsborough is as British as that of Winston Churchill and Rolls Royce. They are each symbolic of British integrity and durability. To older generations the name conjures up the strains of 'Rule Britannia' and an era when the sun never set upon the Empire.

Steam, tea and tractors

When the first Marshall portable steam engine was produced at the Brittania Iron Works, Gainsborough, in 1857, a mere nine years from the time the firm commenced business, neither William Marshall nor his sons James and Henry could have foreseen the future development and status for which the company was destined.

Marshall steam engines gained an international reputation for dependability. The resulting sales meant that the firm faced a continuing requirement to increase the production output of the Gainsborough Works in order to cope with the home and overseas demand. The year 1877 saw the introduction of the Marshall steam traction engine. The cylinder and 'motion' were located above the horizontal boiler. This design concept endured for half a century and was the pattern adopted by the majority of traction engine manufacturers around the world.

In 1857 a Scot from Aberdeen, who was an extensive tea plantation owner in Assam, approached Marshall with a plan to design tea-rolling and processing machinery. By that year the production of tea on plantations in India was valued at 35 million pounds. The firm immediately identified an opportunity for diversification and expansion into lucrative new markets. Machines were soon manufactured and their export commenced. By the end of the 19th century, Marshall-Jackson tea-production machinery was the most advanced in the world and completely dominated the markets in India and Ceylon.

The output from the Britannia Iron Works at the turn of the century had reached monumental proportions. Added to the range of portable steam engines, traction engines and tea-production machinery were boilers, steam rollers, threshing mills, drilling rigs, flax harvesting machinery, concrete mixers, pumps, electric light generators, hay elevators and numerous other lines of agricultural machinery.

In 1905 Marshall contracted Mr Herbert Bamber to design an internal combustion engine that would be suitable for powering an agricultural tractor. Bamber had achieved impressive results in his capacity as senior design engineer at the Luton-based Vauxhall Motor Company. The engine Bamber subsequently designed for Marshall, a 30 h.p. 4-stroke twin-cylinder with a 7 x 7 inch bore and stroke, was ready for testing in 1906. It was mounted in a specially constructed tractor and taken to an adjacent Lincolnshire farm to undergo rigorous tests.

Pulling two 3-furrow mouldboard horse ploughs, the tractor was able to plough one acre in just over an hour. This was considered satisfactory, and plans were drawn up to proceed with the production of a range of Marshall tractors powered by variations of the Bamber engine. They were to be known as Marshall Colonial tractors and initially consisted of four types:

Class C	2-cylinder	30-35 b.h.p. single speed, 1.75 m.p.h.
Class D	4-cylinder	60-70 b.h.p. single speed, 1.75 m.p.h.
Class E	2-cylinder	30-35 b.h.p. dual speed, 1.75 and 3.5 m.p.h.
Class F	4-cylinder	60-70 b.h.p. dual speed, 1.75 and 3.5 m.p.h.

Each of the above was fitted with the 7 x 7 inch bore and stroke cylinders. Class E and F tractors were also available with winding drums (for cable ploughing, winching, etc.) each with the capacity to accommodate 100 yards of either .625 or .75 inch steel rope.

A Class G 4-cylinder tractor was also introduced. It was basically a Class F equipped with spring suspension, designed for road haulage.

The 1914 Marshall Colonial Class F, restored by Queensland farmer Nev Morris, is fitted with the 'Tropical' water cooling system. This tractor is powered by the Bamber-designed 4-cylinder 60-70 b.h.p. engine. (Photo: I.M.J.)

The standard cooling method was with the utilisation of hopper cooling tanks, however 'Tropical' versions could be ordered with fan-cooled tube radiators.

Marshall tractors were purchased by broadacre farmers in Argentina, Australia, Canada, India, Persia and South Africa. They were generally considered overly heavy for application in the soft, moist soils of the fertile British farms, therefore few were retained within the British Isles. During a six-year period (between 1908 and 1914) around 300 were built, nearly all being exported. Certainly they were powerful tractors, but they earned an appalling reputation for breaking connecting rods and did not share the degree of reliability associated with the majority of their American counterparts.

Marshall Colonial tractors in Australia

There are no reliable records of how many Colonials came to Australia but it is generally accepted that the figure is likely to be around 20. It is not surprising that out of the original 300 produced a mere handful remain in existence around the world. (It is interesting to note that Marshall Sons & Company Ltd., in the 1980s, purchased the remnants of a Class F tractor from a collector in Benalla, Victoria, and returned it to the Gainsborough plant. Following 13 months of painstaking work this unit was restored to as-new condition. It is believed to be the only complete and running example of a Colonial in Britain.)

Australia is indeed fortunate to be able to boast possession of at least three Marshall Colonial tractors in good going order, plus one or two others in the process of being restored.

Queenslander Nev Morris displays his impressive Class F Tropical each year to enthralled spectators at the annual Jondaryan Woolshed Rally. A second Class F is part of a Western Australian collection. A Class E at Victoria's Swan Hill Pioneer Settlement is the most recent Marshall Colonial to have had its restoration completed — and that is a story in itself!

The saga of the Swan Hill Marshall Class E

Clare Station occupied an area that would have encompassed several English shires. It lay a three-day drive for a bullock team, south west from the remote outback town of Ivanhoe, in western New South Wales. It was to that lonely place around 1911 that a brand new Marshall Colonial Class E tractor was delivered. It had travelled half way around the world, lashed to the deck of a sailing clipper, since leaving the soft greenery of its place of birth, Lincolnshire.

One can only speculate how it was finally transported to Clare Station. It almost certainly was shipped to Adelaide then transferred to a smoke-belching paddle steamer for its thousand mile trip up the Murray Darling to Menindee. Following the two-month river trip it would have been uncrated and prepared so that it could continue on the final leg of the journey under its own power. This would have involved several days of clattering through semi desert country — the precise route being determined by the proximity of waterholes and avoidance of eroded gullies.

Its duties at Clare Station were twofold: the powering of the overhead gear during the shearing and crutching seasons and the hauling of wool to the river port (and later the railhead), returning with station supplies.

Little is known of the intervening fortunes of the tractor until it was acquired for posterity by the Pioneer Settlement museum village at Swan Hill, Victoria, around 1964. There, a new chapter was to unfold for this now historic and rare Marshall Colonial. Tractor enthusiasts, historians and cheerful tourists would be invited to marvel at the magnificence of this impressive example of early British engineering.

Sadly though, attempts to restore the Marshall were impeded by a lack of adequate funding and (it would appear) expertise. There were three separate attempts to return it to its original glory. It was not until 1990 that Newton Williams, an engineer and professional classic tractor restorer of considerable ability, was requested by the new management of Pioneer Settlement to take the big tractor in hand and carry out a 'preventative maintenance programme'. Technically, the funding could not be made available for a 'restoration' as officially there had been *three* restorations performed since 1964. The resourceful Williams interpreted his assignment as a requirement to start at the front axle and move to the rear drawbar, *fixing everything in between.*

Newton Williams recalled that when he had last seen the engine started it had required all its power to move the 20 000 lb. tractor just a few feet. The reason for this immediately became apparent when the front axle was jacked up and it was found that the wheels were seized on the axle through lack of grease.

When the engine did start (as distinct from 'run') a sound resembling powerful blows from a 12 lb. sledge hammer emitted from within. There was no alternative, therefore, but to face the somewhat daunting task of removing the 1-ton engine from the frame for the purpose of proceeding with exploratory surgery.

The first job after the crane had lowered the engine onto the concrete floor was to remove the crankcase inspection covers. This revealed handfuls of white metal — discouraging to say the least. Obviously the bearings had melted due to lack of lubrication.

The jackets of the cylinder heads, which were in a similar state of deterioration, were re-fashioned by using cast iron pieces salvaged from early Swan Hill street lamps. They were welded into place with the assistance of a cast iron metal spray — all of which was a learning experience for Newton. 'Just imagine', quipped Newton, 'a cylinder head too heavy to lift, that has to be heated and cooled *very* slowly, and then to hear it go crack — as you are washing your hands when cleaning up!'

The crankshaft had to come out. This revealed that an earlier restorer had left *stillson marks on the rear bearing end of the shaft*. The decision was taken to send the crankshaft to an engineering firm for repair. What followed is almost beyond belief. The big ends were destroyed in the 'repair' and the bronzing of the bearing surface of the crankshaft ends could be scratched off with a finger nail! Over a period of three-and-a-half years the crankshaft was sent to no fewer than *four* expert repair firms.

The big 2-cylinder engine being lowered into the frame of the tractor, having been completely rebuilt. (Photo:N.Williams)

Finally, on the day when the crankshaft was ready for fitting back into the engine, it was discovered that a set of essential dowels and special bolts had been lost by the last of the crankshaft repairers. They claimed they had never received them with the shaft. New dowels had to be manufactured by Newton. Not a simple task, as there were no examples from which to copy. Following hours of precision lathe work, the new dowels and matching bolts were ready. Ten minutes later the originals turned up in the mail!

The remaining work on the Marshall was temporarily put aside as other pressing jobs required attention, but one week prior to the 1995 Swan Hill Centenary Parade, the manager of the Pioneer Settlement mentioned casually to Newton Williams that he wanted the Marshall to be ready for 'next Friday's parade'!

The following seven days, according to Newton Williams, went something like this:

> Monday morning crane arrived to lift engine, also radiator tank, on to tractor frame — find parts — assemble and fit — machine up new clutch release bearings and fit (due to geometry this has to be done *in situ*) — modify clutch catch — find vent pipe for gearbox — found — repair the damage to it caused when the kids kicked it off — repair the roof — fit new timbers — find all the plumbing for the oilers and fit — organise painting and signwriter — answer the phone — eat occasionally — sleep (ha ha).

> Thursday — fuel up and start. After oiling up — prime cylinder with fuel through cocks on cylinder heads and press button. Well — that's the theory, but as there are parts missing from the electrical department it has to be tow started. It starts! 'Don't stop it!' Drive around the Settlement — everyone comes out to gaze. Even the C.E.O. of the River City of Swan Hill hangs out from his upstairs window to wave, with a satisfied expression upon his countenance suggesting that it was all *his* own work.

Friday dawned bright and sunny, a great day. The engine fired up instantly and with young Mark Smith recruited to be helmsman, the big machine was manoeuvred into position behind the police car which initially led the parade. Mark's main problem was that visibility dead ahead of the Marshall was totally obscured by the large square radiator tank. In between tinkering with the carburettor and checking the multitude of adjustments, Newton Williams endeavoured to navigate by peering around the radiator tank and issuing frantic instructions to Mark. The problem was that the police car insisted on remaining a mere two metres in front of the clattering tractor. Mark had only had his licence for a few weeks and was visualising how he would explain to the magistrate why he had demolished a perfectly good police car whilst travelling at only 3.75 miles per hour.

The Marshall Class E Colonial tractor majestically leading the Swan Hill 1995 Centenary Parade through the town. (Photo: Gillian Williams)

The cheering crowd loved the big Marshall. It was undoubtedly the highlight of the Centenary Parade. Was all the effort and expense worth while? According to Newton Williams — every bit of it.

The big Marshall Colonial Class E now languishes (somewhat smugly perhaps) in its prime location at the Swan Hill Pioneer Settlement. Undoubtedly, there it will remain for all time, for future generations yet unborn to enjoy its magnificence and perhaps reflect upon its past.

The restored Marshall Class E may be inspected at the Swan Hill Pioneer Village, on the Horseshoe Bend of the Murray River, Victoria.
(Photo: N. Williams)

Later Marshall tractors

A 1938 Marshall Model M, restored by Canadian Marshall authority Stan Kick. The M was the fourth version of the single-cylinder 2-stroke diesel engined tractors first produced in 1930. It was also the most successful and was the tractor upon which the postwar Field Marshall tractors were based.
(Photo: Stan Kick, Alberta)

The 1949 Field Marshall Series 3 utilised the same 6.5 x 9 inch bore and stroke engine as the Model M, but incorporated a two ratio gearbox producing 6 forward and 2 reverse speeds. The tractor was popular in Australia and New Zealand and sold in direct competition to the semi diesel Lanz Bulldogs. The tractor pictured is owned by Mal Cameron of South West Rocks, N.S.W. and was photographed during the 1996 Rusty Iron Rally. (Photo: I.M.J.)

1957 Marshall MP 6. Powered by a Leyland UE 350 6-cylinder diesel engine developing 68.6 b.h.p., this was one of the most powerful British wheeled tractors of the era. Its relatively high price of £1475 restricted its sales. (A David Brown 50D could be purchased for £891 and a Massey Harris 745 for £725.) (From an original Marshall sales brochure)

1971 Track Marshall 90. Marshall chose the popular Perkins 6/354 6-cylinder diesel engine to power the 90, giving it a maximum sustained drawbar pull of a respectable 15 400 lbs at 1.9 m.p.h. The operational weight of nearly 7 tons enabled the 82 b.h.p. to be transferred to the ground with minimal track slip. (From an original Marshall sales brochure)

1982 Marshall 554. In 1979 a Lincolnshire farmer and financier named Charles J. Nickerson purchased a portion of the Britannia Iron Works and the rights to the production of Track Marshall tractors. The Marshall Company had been taken over by the Thomas W. Ward Group in 1968, who in turn sold their interest to British Leyland in 1975. The firm was renamed Aveling Marshall Ltd. In 1982 Nickerson acquired the Leyland Tractor Division from Leyland Vehicles Ltd. The Marshall 554 4 w.d. wheeled tractor evolved as a result of this acquisition. The new firm was re-registered as Marshall Sons & Company. A complex series of acquisitions indeed! (Photo: I.M.J., by kind permission of Lincolnshire Museum of Country Life, Lincoln. U.K.)

CHAPTER 22

Early Tractors in Scotland

Scotland is not just a nation of kilted, hairy Highlanders living on remote crofts, perpetually playing the bagpipes and quaffing rare malts, as some tourist brochures might suggest. It is a land rich in agricultural heritage.

This chapter examines the evolution of mechanisation on Scottish farms, which are now rated as being among the most productive in the world. (An abridged version of this text first appeared in *Australian Grain* magazine in 1996.)

First — the geology lesson

An agriculturalist visiting the United Kingdom might well be surprised to discover that some of the most productive arable land is to be found in Scotland. There is an interesting geological explanation for this which is worth a brief examination.

During the last Ice Age the massive southerly assault of a creeping glacier halted on a line roughly extending between the Firth of Clyde on the west coast and the east coast town of Stonehaven, which lies 30 kilometres south of Aberdeen. Pushed ahead of the glacier had been untold thousands of cubic kilometres of productive silt scoured from Scotland's north. When the glacier finally melted it left behind this bounty of nutritious soil, rich in trace elements and humus.

As a consequence of this primeval activity, the topography of mainland Scotland is today divided into three distinct regions. The Highlands in the north are predominately mountainous, with steep peaks and deep ravines gouged out by the glacier and not yet mellowed into roundness by age. The great thaw left the ravines brimming with water, thus forming the incredibly beautiful lochs. The Lowlands in the south were unaffected by the last Ice Age and are therefore mellowed and softened with greater antiquity, but lack the scenic grandeur and elevation of the younger Highlands.

There remains between the Highlands and Lowlands an area known in geological terms as the Central Lowland Rift — the legacy of the rich silt left behind by the glacier. It embraces what are now the fertile farmlands of the Lothians, Fife, Dumfries and Berwick, and extend into Perthshire. These relatively flat to undulating areas, with their lush deep top soil, have been farmed for thousands of years — long before the Romans unsuccessfully attempted to conquer the land from its fiercely protective clansmen.

Scottish farmers unimpressed

During the latter half of the 19th century *steam plough engines* were used with only marginal success on some of the larger English estates. One of these giant rigs was tried out on Fenton Barns, a prominent Scottish grain farm in East Lothian. The capital cost involved was eight hundred pounds and, in relation to its productivity, the steam machine was a flop. Neighbouring farmers witnessed the 12-ton steam plough engine becoming frequently bogged in the soft, moist loam soil and came to the opinion that the horse would never be replaced by a mechanical device. The steamers were considered suitable only for powering threshing mills, grinders and pumps.

A rural scene depicting cable ploughing using a steam plough engine. (From an original Fowler sales brochure, courtesy W. Johnson)

Whilst the American, Australian, New Zealand, English and European farmers experimented with tractors in the early days of the new century, Scottish farmers were singularly unimpressed. Certainly, this may well indicate a lack of vision, but it just possibly could denote a considerable degree of shrewdness on their part. Early tractors were expensive, unreliable, inefficient and dangerous.

Even the diminutive Ivel tractor, manufactured in Bedfordshire by Dan Albone, did not excite the Scots. This was despite the fact that it displayed a reasonable degree of reliability and by 1908 was being exported to countries as far as Australia. In 1904 a farm machinery dealer in Perthshire had demonstrated an early production model of Ivel to a group of curious local farmers, but their hard-earned money remained resolutely in their sporrans.

The drive for increased production

During World War 1 Scottish farmers were urged to open up new land to the plough and the Minister for Scottish Affairs wanted crop production doubled. Responding to the appeal, the Royal Highland Agricultural Society arranged a series of tractor field days, the first being held in 1915 on a farm near Stirling. The turnaround in interest was dramatic. Two years later, in 1917, a major demonstration was arranged at Gragie Hall outside Edinburgh, resulting in scores of tractors being sold.

The first tractor to be sold in volume in Scotland was the Fordson F manufactured at Dearborn, Michigan. It was powered by a 4-cylinder 20 h.p engine. (Photo, circa 1926, courtesy Hugh Lemon, Newtownards, depicts his father ploughing in County Down, Northern Ireland where, as in Scotland, the Fordson F was the top selling tractor.)

In that year John Wallace & Sons of Glasgow commenced the importation of Moline tractors from Illinois. Meanwhile, the British government made a special deal with Henry Ford and ordered a staggering 6000 of the new Fordson Model F tractors to be shipped from Dearborn, Michigan. Many of these were snapped up by the now enthusiastic Scottish farmers. Deere & Co., also of Illinois, sent Waterloo Boy tractors across the Atlantic, where they were sold under the curious but significant name of 'Overtime'. All these tractors were relatively compact and light in weight, suiting the soft Scottish soil.

Guthrie's Glasgow

In 1918 a historic meeting took place between the Scottish industrial firms of John Wallace & Sons, the Carmuir Iron Foundry and the D.L. Motors Manufacturing Co. These well regarded firms agreed to pool their resources and produce an indigenous Scottish tractor. Following some prototype developmental work at premises in Motherwell the new company moved into a disused munitions factory at Cardonald, on the outskirts of Glasgow.

The tractor was designed by the brilliant Scottish engineer William Guthrie and named the 'Glasgow'. The syndicate drew up a set of challenging specifications for Guthrie, the most important of which was that the new tractor had to be capable of working continuously in the soft, sticky soils of the Lothians without the problems of bogging. Bogging was a major argument against tractors in Scotland. A frequent occurrence involving tractors was the necessity to 'louse' a horse team from an adjacent field to pull a hopelessly bogged tractor onto firm ground.

Guthrie tackled the problem in a novel way. He designed the tractor as an *all-wheel-drive* three-wheeled tractor. It featured two front wheels and a third centrally located rear wheel. In effect, no wheel followed in the track of another. Each time the tractor moved forward it was propelled by three wheels all operating on fresh ground. There was no differential in the front axle, therefore a single wheel could not slip or spin. All three wheels would have to spin together — an unlikely event. In place of a differential the front wheel hubs were fitted with a pawl and ratchet system. This permitted smooth turning by allowing the outside wheel to speed up whilst the other continued at the sustained speed in relation to the rear wheel. Interestingly though, in reverse gear the ratchets disengaged the front wheel drive so the Glasgow became a *one-wheel-drive*!

Extract from a Glasgow sales brochure of 1920.

Cost of Ploughing

Careful field tests made have shown that the GLASGOW TRACTOR can plough 1¼ acres of medium quality land per hour, working this at a depth of 8½" to 9" in stubble land, and 6" to 7" in lea land with a consumption of 1¼ gallons per acre.

✤ ✤ ✤

THE COST PER ACRE can therefore be determined as under :—

Under *favourable* climatic conditions the acreage ploughed per week of 5 days of 8 hours each would be 5 × 8 × 1¼ = 50 acres

The cost would be :—

Petrol, 1¼ gallons per acre @ 3/- per gallon = 1¼ × 3/- × 50	£13 2 6
Operator's Wage	3 0 0
Depreciation (assumed at 25 per cent. per annum if Tractor ran continuously) per week	2 0 0
Oil and Repairs and Sundries	1 0 0
	£19 2 6

Cost per acre ploughed— £19 2s. 6d. ÷ 50 = 7s. 8d. per acre.

COST PER ACRE—continued

Under *unfavourable* conditions assume that only 2½ full days are worked, the acreage falling to 25 per week

The cost would then be :—

Petrol, 1¼ gallons per acre @ 3/- per gallon = 1¼ × 3/- × 25	£6 11 3
Operator's Wage	3 0 0
Depreciation, as before; but for shorter time	1 0 0
Oil, Repairs, „ „ „	0 10 0
	£11 1 3

Cost per acre ploughed— £11 1s. 3d. ÷ 25 = 9s. per acre.

✤ ✤ ✤

THESE tests also proved that the Tractor will climb hills which are at present almost too steep for horses (*i.e.*, gradients of 1 in 5) and will pull a three-furrow plough in medium soil working at a depth of 6" to 7".

✤ ✤ ✤

COMPARE this with horse ploughing and note the difference.

EARLY TRACTORS IN SCOTLAND 165

The Glasgow 3-wheel-drive was a brilliantly engineered tractor powered by a 27 h.p. Waukesha 4-cylinder engine, although some units were equipped with Burt-McCollum or Continental power plants. (Photos: Newton Williams, Swan Hill Pioneer Settlement)

The Glasgow was rushed south of the border to the Lincoln Tractor Trials in 1919. Competing against British, North American and Italian tractors it recorded the highest pull-to-weight ratio of any tractor, equivalent to 76% of its gross weight.

The schematic drawing of the Glasgow clearly shows the design of the drive train to the three wheels. (From an original technical brochure)

After such great promise, the Glasgow tractor tragically went out of production in 1924. This was entirely due to the sudden financial collapse of its principal distributor, the English firm of British Motor Trading Corp. Ltd.

Tractors come of age

By 1925 there were approximately 1700 tractors operating in Scotland. The most popular was the Fordson Model F, largely because of its price advantage. There were, however, other American tractors making inroads into the hearts (or wallets) of Scottish farmers, including various models of Internationals. There was also the Parrett tractor from Chicago which had shrewdly been renamed 'Clydesdale'. This, not surprisingly, attracted the interest of some of the horse diehards.

The Parrett tractor was sold in Scotland as a 'Clydesdale' in an endeavour to attract the interest of those farmers who still believed in the invincibility of the horse. (Photo I.M.J., courtesy Western Development Museum, North Battleford, Saskatchewan)

According to the Scottish Department of Agriculture, at the outbreak of World War II in September 1939 there were around 7000 tractors ploughing up Scottish farms. But it was not until the latter 1940s with the advent of firstly the Fordson E27N Major, closely followed by the David Brown Cropmaster and then the grey Ferguson, that Scottish farmers finally accepted that the days of the draught horse were coming to an end. By 1960 the farm horse in Scotland had been relegated to an item of nostalgia and curiosity.

The 1954 and 1956 Scottish Farm Machinery Census

The following table reveals the interesting evolutionary trend of Scottish mechanised farming during the industrious mid-1950s. The figures relate to the actual number of the various machines in regular use on Scottish farms.

The 'Dain' John Deere entered production in 1916. It was equipped with an ingenious on-the-go gear shift arrangement. Each of the three wheels were drivers, thus providing excellent traction. An estimated 200 were built and, amazingly, one fully operational example remains today (production no.79) in the caring hands of a Minnesota collector. (Photo, circa 1918, with special permission from Deere & Co., Moline)

With a degree of urgency the board looked around for an alternative. They focused their attention upon the Waterloo Gasoline Engine Company which had been manufacturing Waterloo Boy tractors at its plant in Waterloo, Iowa since 1913. H.W. Leavit, the firm's chief engineer, originally designed a tractor named the Big Chief for a rival company. Upon joining the Waterloo Gasoline Engine Company, Leavit had simply redesigned the Big Chief, introducing numerous improvements, and thus created the Waterloo Boy. In 1918 Deere & Co. purchased the Waterloo Gasoline Engine Company, together with the Waterloo Boy tractor and the services of H.W. Leavit. By this time over 8000 of the current model twin-cylinder Waterloo Boy tractors (the Model N) had been purchased by farmers around the world, who were more than pleased with their new tractor. Some were exported to Britain, where they were marketed as the Overtime Model N. (Interestingly, Ulsterman Harry Ferguson become their distributor in Ireland.)

Despite the change of ownership, it was decided to continue to sell the tractor under the brand name of Waterloo Boy until 1923, when the John Deere Model D was released. Significantly, around six years had elapsed since Dain's 3-wheeled tractor went out of production.

This 1920 photo is of a Waterloo Boy Model N pulling a set of tandem disc harrows in a corn field somewhere in Illinois. (With special permission from Deere & Co., Moline)

This studio photograph of a Waterloo Boy Model N was taken on 8 June 1920 and subsequently used for advertising brochures. Unlike most tractors of the period, the belt pulley of the Waterloo Boy was mounted well clear of any obstructions on the front of the tractor and therefore much favoured for belt operations. (With special permission from Deere & Co., Moline)

The Model D

The twin-cylinder John Deere Model D became a legend. It was commercially one of the most successful tractors of all time. It also had the longest production life of *any* model of *any* tractor. Including several upgrades, it was produced from 1923 until 1953. In this period, nearly 200 000 were manufactured.

During the 30-year production life of the Model D, and up to 1960, a comprehensive range of John Deere tractors was produced. This included row crop, broadacre and orchard tractors also crawlers and industrial units. Each in its own way was a classic; each in recent years has become much prized by collectors and museums around the world.

The concept of the original brilliantly simple Waterloo Boy horizontal 2-cylinder water-cooled engine was retained until 1960. It, of course, was given numerous capacities and refinements over the years, including diesel variations introduced in 1954.

John Deere Model D 1931. This was one of the most successful tractors ever, having a production life of 30 years. (Photo: I.M.J., with kind permission from Peel Valley Machinery Co., Tamworth, N.S.W.)

A new era of tractors

In 1960 Deere & Co. unveiled an all-new range of tractors involving major design changes. The new 10 Series incorporated the very latest agricultural tractor technology. Gone were the faithful twin-cylinder engines, replaced now by modern multi-cylinder diesel power units. A mere 5% of components were carried over from the earlier tractors. The new machines had been developed in great secrecy and caused a furore when they were released.

From 1960 up to the present day, there has been conspicuous evidence of the global success and development of John Deere farm tractors. Research and development receives the highest priority, ensuring that the company's remarkable achievements, since John Deere produced his steel plough in 1837, will continue into the next millennium.

It is interesting to ponder how tractors have developed during the 20th century. What would Joseph Dane think if he could experience a 1996 John Deere 8200? (From an advertising brochure courtesy John Deere, Australia)

CHAPTER 24

The *Motor* Cable Ploughing Engines

Mr W.L. (Bill) Johnson, former Director and General Manager, J.C.B. Australia Pty.Ltd., was the son of a prominent Norfolk farmer. He grew up during the fascinating and little known era of the motor cable ploughing engines (as distinct from *steam* ploughing engines). I am privileged to have been entrusted with historic photographs and documents from the Johnson family album for reproduction in this work. But even more precious and significant are the first-hand accounts related to me by Bill Johnson of these long gone days, when hard physical work on the land was considered a virtue.

The chapter begins with a brief reference to John Fowler's steamers, as it was from these hissing, belching machines that the internal combustion powered *motor ploughing engines* evolved.

John Fowler

John Fowler was born in 1826 in the village of Melksham, deep in the picturesque Wiltshire farming country. At the early age of 20, already showing considerable agricultural engineering talent, he took an interest in the problems facing the Irish peasants during the Great Potato Famine of 1846. Fowler believed the salvation of the Irish economy would be to drain the vast areas of unproductive bogland, thus enabling them to be converted into fertile arable farming country. He envisaged a type of large-diameter deep mole drainer that could create an underground drainage system.

In 1850 Fowler demonstrated at Exeter, before a committee of the Royal Agricultural Society of England (R.A.S.E), a mole drainer that was able to lay a drain 2 ft. 6 inches deep in tight clay. The drainer, attached to a cable, was pulled through the ground by a pair of draught horses winding a capstan. Three years later he was successful in increasing the working depth of a new design of mole drainer, which required four horses. The slow work was punishing for the horses and he felt there had to be a better method of reclaiming new ground. John Fowler turned his attention to steam engines and cable-drawn ploughs.

An early Fowler cable-drawn mole drainer. Note the cutaway drawing showing the mole in the ground. (From an original catalogue, courtesy W.L. Johnson)

The Fowler Steamers

Fowler built his first 'set' of ploughing engine tackle in 1856, using a Ransomes & Sims portable steam engine, and demonstrated it to a meeting of the R.A.S.E. at Chelmsford. The committee dismissed the apparatus as inefficient, owing to the cost of ploughing one acre being seven shillings and tuppence halfpenny. It was claimed that a team of horses could achieve the same result for a mere seven shillings! (Apparently no consideration was given to the fact that the steam equipment could work non-stop during daylight hours.)

In 1857 the Royal Highland Agricultural Society of Scotland recognised the potential of steam ploughing and awarded Fowler a prize of two hundred guineas for his endeavours.

A rural scene depicting steam ploughing in action using a pair (or, more correctly, a 'set') of 2 steam ploughing engines and a plough. (From an early engraving, courtesy W.L. Johnson)

In 1886 John Fowler joined forces with a financier named William Hewitson and created a limited company, which was later to evolve into John Fowler & Co. (Leeds) Ltd. and become one of England's largest and most prestigious manufacturers of steam engines. By 1910 there were 650 sets of Fowler steam ploughing tackle working on British farms.

A Fowler BB compound cable ploughing engine, circa 1918. (From an original catalogue, courtesy W.L. Johnson)

As the farm tractor and trailed plough gradually evolved into reliable and affordable items of agricultural machinery, most farmers considered them a better alternative to steam cable ploughing. However, the introduction of petrol and diesel engined cable ploughing tractors extended the era of the cable plough for around a decade. So, as the internal combustion powered tractors developed, the steamers declined in popularity, until in 1933 the last set of Fowler steam ploughing engines was sold to a plantation owner in Formosa (Taiwan).

(Note. The saga of the steam traction engines is an immense topic about which numerous books have been published. As *this* book essentially deals with the internal combustion powered tractors, only a brief reference is given to the subject of steam in order to highlight its relevance to the progression of the farm tractor.)

Hercules and Samson

On 7 July 1927 Mr Arthur Reginald Johnson of Bentinck Farm, Terrington, Norfolk, received an invoice from John Fowler & Co (Leeds) Ltd. indicating that his set of motor ploughing engines was ready for delivery.

For as long as there have been farmers, there have been those who have especially excelled at their vocation, thus being elevated in the estimation of their peers to a position of respect and prominence. Their expertise is almost always due to a successful infusion of aptitude, hard work, affinity with the soil and ability to recognise better methods of farming technics. Mr. Johnson qualified for the foregoing credentials and was indeed a competent agriculturist, as evidenced by the considerable acreages he was able to accumulate. His innovative farming methods frequently served as an inspiration to his neighbours.

Raised on the land, Mr A.R. Johnson grew up in a world of Belgium, Suffolk Punch, Percheron and Shire draught horses. He had also experienced the early days of steam and all its accompanying complications. A major problem for the owners of steam engines in his local Norfolk area was the quality of available water. Steamers had a rapacious thirst and, if fed with hard chemical-contaminated water, the resultant adverse effect on the boiler and steam valves could be catastrophic.

Accordingly, in 1927 farmer Johnson saw considerable merit in investing in a set of Fowler petrol engined cable ploughing tackles. The set consisted of:

2   150 h.p. tractors fully equipped for cable ploughing
1   dormitory and kitchen van
1   5-6 furrow anti-balanced plough
1   11-13 tyne cultivator/beet lifter
1   3 furrow zig zag frame balanced plough

The discounted price of £5285.15.10 included rail freight from Leeds to Peterborough per the London & Northern Eastern Railway.

In fact, the two tractors, which Mr Johnson named 'Hercules' and 'Samson', were not identical. Samson (tractor No. UA528) was the left hand engine and Hercules (tractor No. UA529) operated on the right. Their respective winches worked on opposite sides, enabling the tractors to work in tandem at either side of the field.

1927 Fowler No. UA528 motor cable ploughing engine. The driver is Mr Albert Reynolds, the wee boy is the owner's son, Bill Johnson, and standing alongside is the owner, Mr Arthur R. Johnson. The tractor was known as Samson and stood on the right hand headland with its cable feeding to the right.
(Photo, circa 1929, courtesy W.L. Johnson)

The original documentation forwarded to Mr A.R. Johnson by Fowler is reproduced on the following pages. Despite having suffered the ravages of age, they represent an interesting historical document

TEAM PLOUGH & LOCOMOTIVE WORKS.
LEEDS.

MELUN: 2, RUE DE LA VARENNE.
BOMBAY: FOWLER BUILDING.
HAVANA, REPUBLICA DEL BRASIL,II.
MANILA, P.I. P.O. BOX 1064.
PRAHA-LIBEN, VIS A VIS STAATSBAHNHOF.
BUDAPEST, KELENFÖLD: VIS A VIS BAHNHOF.

113, CANNON STREET,
LONDON, 7th July 1927.
E.C.4.

A. R. Johnson Esq.,
Bentwick Farm,
Terrington St. Clements,
Kings Lynn.

To John Fowler & Co (Leeds) Ltd

REMITTANCES: ALL CHEQUES, DFTS. &C. SHOULD BE SENT TO 113, CANNON ST, LONDON.
ORDERS: ALL ORDERS SHOULD GO DIRECT TO STEAM PLOUGH WORKS, LEEDS.

N.B. SHOULD ANY OF THESE GOODS BE RETURNED, PARTICULARS OF SAME MUST BE SENT AT THE TIME OF DESPATCH TO JOHN FOWLER & C°(LEEDS) L° STEAM PLOUGH WORKS, LEEDS, OR OTHERWISE NO CREDIT WILL BE GIVEN.

INTEREST AT 5% CHARGED ON OVERDUE A/CS.   TERMS

1927 July 7.	NOS. 17536/7. TWO. 150 H.P. MOTOR CABLE PLOUGHING ENGINES, Each with:- governors, three ploughing speeds: three road speed forward and three reverse: Steel winding drums with self-acting coiling gear & provided with 500 Yards of Fowler's Special Steel Wire Rope: vice on tender: hind road wheels 6'-6" dia. x 22" wide: front road wheels 5'-0½" dia. x 14" wide, belt pulley 24" dia. x 7" wide, tank capacity 85 Gallons, Brass nameplate:- A. R. JOHNSON TERRINGTON ST. CLEMENTS. SAMSON & HERCULES. Electric horn, Steel toolbox, etc., and outfit complete. EXTRA FOR:- Rotary Pumps. Awnings. Electric self-starting gear. NO. 14112. ONE. STANDARD SLEEPING VAN COMPLETE WITH Beds and Bedding for five men, fitted with cupboards, desk, stove & chimney, & simple cooking utensils, dragbar, vice etc., complete.			
		4250	–	–
		5	–	–
		38	–	–
		182	–	–
		278	–	–
	FORWARD.	4753	–	–

(2)
R. Johnson Esq.,

BROUGHT FORWARD.   4753 - -

No. 14616.
ONE. 5/6 FURROW ENGLISH TOPSOIL ANTI
BALANCE PLOUGH, fitted with 12 Steel skifes
G.123/4, Steel mouldboards No.20, shares,
sleds, & coulters, furrow wheel 5'-6" diam. x
8" wide Land wheel 5'-0" x 9" plus 3",
drum skid wheels 24" x 5", steel toolbox
containing:- 7 spanners, 1 Ratchet &
handle, 2 Spare shackles, & spare bolts
& nuts.   336 - -

No. 14566.
ONE. 11/13 TYNE COMBINED CULTIVATOR &
BEETLIFTER, fitted with Steel tynes & points,
double sockets for back tynes, Patent
cushioning cylinder, back drag bar,
wooden toolbox containing:- 7 Spanners,
1 Quarter hammer & shaft, 1. ¾" Box
spanner, 1 Ring spanner for sandcaps,
1 Wedge drift, 1 Set of Beetlifter tynes
& shares.   383 - -
EXTRA FOR:-
SELF-ACTING lifting gear.   30 - -

No. 14287.
ONE. 3 FURROW ZIG ZAG FRAME BALANCE PLOUGH,
fitted with 6 Steel skifes C49/50, Steel
mouldboards No.194, Steel plate shares, sleds
& 2 coulters, six Subsoil tynes and points,
Furrow wheel 5'-6" x 8", Land wheel 5'-0" x
9" plus 3", skid wheel 24" x 5", Steel
Toolbox containing:- 7 spanners, 1 Ratchet
& Handle, 2 Spare shackles & pins & spare
bolts & nuts.   342 - -
EXTRA FOR:-
Subsoil tynes, Sockets, Points etc., and
heavy skifes for taking same.   29 2 -
                                       371 2
                                 £   5873 2 -

Per L.& N.E.Rly. to Peterborough, for exhibition
at The Peterborough Show, to be delivered direct
to you at close of show.
Carriage paid.

STATEMENT.

113, CANNON STREET,

LONDON, 7th July 19 27.
E.C.4.

A.R. Johnson Esq.,

Bentwick Farm, Terrington St. Clements,
King's Lynn.

To JOHN FOWLER & C° (LEEDS) L.TD

REMITTANCES: *Please send all Cheques Dfts. &c. to 113. Cannon St. LONDON, E.C.4.*
ORDERS: *All Orders should go direct to Steam Plough Works, LEEDS.*

1927				
Jly 7	To Motor Cable ploughing Engines Nos. 17536/7 etc. as per invoice.	5873	2	–
	Less 10% discount	587	6	2
	NET	£5285	15	10

13th August /27

RECEIVED

By cheque £5285. 15. 10.

For JOHN FOWLER & CO. (LEEDS) Ltd.

1927 Fowler No. UA529 motor cable ploughing engine with Mr Reg Atkins at the wheel and Mr A.R. Johnson standing by the left feed winch. (Photo, circa 1929, courtesy W.L. Johnson)

Mr A.R. Johnson purchased a second set of motor ploughing engines in late 1927. The sets were used largely for contract ploughing, for which a new company was created called Cable Cultivating Contractors Limited and based at Long Sutton, Lincolnshire. The firm was eventually acquired by Fowler, for whom it served as a convenient means for testing and establishing new equipment.

Balanced and anti-balanced ploughs

The original ploughs designed specifically for cable ploughing were termed *balanced*. As a balanced plough was drawn across a field the mouldboards or discs at the 'rear' engaged the soil. Upon arriving at the headland, the plough was prepared for the haul back across the field. The initial jerk caused the 'other end' of the plough to drop down and engage. Being a two-way plough the furrow was always turned in the one direction. The *balanced* plough had a fixed axle and as a consequence was extremely prone to jumping out of the ground.

Later ploughs were known as *anti-balanced* and were fitted with a sliding axle which moved forward or aft depending upon the direction being pulled. The *anti-balanced* plough kept its mouldboards or discs level and *in* the ground.

At the commencement of opening up a new job the two plough wheels were kept level until the initial furrows were opened. On the return and subsequent crossings the furrow plough wheel was set lower than the land wheel. The skill was to maintain a level *bottom* at all times. This involved precise trimming of the *land wheels* and *furrow wheels*.

A team of five men was employed with each set, consisting of the two tractormen, the ploughman, a foreman and a cook. The ploughman undoubtedly had the least enviable job, as he was obliged to sit on the hard pan seat, often engulfed in dust and always suffering extreme discomfort from the rough ride.

A Fowler 7-disc 2 way anti-balanced plough being winched up by a steamer. Perched on the rear is the foreman, who is making certain that the job runs efficiently. Not evident in the picture is the horse-drawn water wagon that would be kept busy carting water to the two steamers. (From an original Fowler catalogue, courtesy W.L. Johnson)

A Fowler cable turning cultivator. Working width 11 ft., depth of cultivation 14 inches. (Photo I.M.J., courtesy Museum of Lincolnshire Country Life)

The power plants

Fowler motor cable ploughing engines were custom built to suit individual buyers and applications. In 1920 a set of two tractors was demonstrated at Lincoln. These units were powered by 4-cylinder White & Poppe petrol engines. The initial two tractors acquired by Mr A.R. Johnson were equipped with Ricardo 150 h.p. petrol engines with three forward and three reverse road gears. The cable winder was also provided with three speed ranges. Later units were fitted initially with Fowler Cooper diesels and then with 80 h.p. Freeman Sanders 6-cylinder diesel engines. Some extra heavy duty models were powered by a massive 170 h.p. diesel engine procured from M.A.N. of Nuremberg, Germany.

It is interesting to reflect that by the late 1920s the American tractor manufacturers, who had pioneered the trundling prairie giants, had largely abandoned the concept of heavyweight machines. Yet in England, Fowler was producing a varied range of quite outstanding gargantuan units, albeit in limited numbers and usually for cable ploughing.

It is also intriguing to consider that Fowler's main competitor, William Marshall Sons & Co., ceased manufacturing its heavyweight Colonial tractors in 1914 and did not re-emerge with a production farm tractor until 1930, when it introduced the single-cylinder 15-30.

For the record, it should be stated that the era of the motor cable ploughing engines was relatively short-lived. By the mid-1930s they had largely disappeared from the landscape. Sensing their demise, Fowler introduced a range of Gyrotiller attachments fitted to tractors ranging from 30 to 170 h.p. Around 70 units were produced before the initial enthusiasm expressed by agricultural contractors soon evaporated when it became evident that the expensive rotary tilling machines were virtually self-destructive and simply not suited to the moist clay British soils. Like the cable plough, the Gyrotiller disappeared from the rural scene prior to World War II.

A Fowler crawler Gyrotiller in big trouble. Bogging could be a problem for these heavyweight machines. Unlike the cable ploughing engines, which remained on the comparatively firm headlands, the Gyrotillers were obliged to criss-cross the paddocks and sometimes became stuck in mud. To extract the bogged unit pictured would be a major headache as there would be few machines with enough horsepower to tackle the job. Note that the Gyrotiller cultivated the soil with a similar action to an egg beater, as distinct from the lateral rotary action of the modern rotary hoe. (Photo courtesy W.L. Johnson)

McLaren's 'Diesel Oil Engine Cable Tackle'

J.H. McLaren Ltd. established the Midland Engine Works in Leeds in 1876. In many ways the firm replicated the progress of John Fowler. McLaren steam engines were held in high regard throughout the world. They were introduced into Australia in 1890, including a number rigged as cable ploughing engines.

Following the return to normality after the 1914-18 war, the McLaren board conjectured that the days of steam power on the land were coming to an end and would be replaced by internal combustion engines. This was a view shared by Fowler, but the two philosophies differed somewhat. McLaren resolutely believed that the future lay in oil (diesel) powered engines, whereas Fowler at that stage was undecided between compression ignition diesel fuelled and low-compression petrol fuelled engines. As a consequence, McLaren pushed ahead with drawings for a tractor powered by a diesel engine and fitted with a cable winch. It was referred to as the *McLaren Diesel Oil Engine Cable Tackle.*

The McLaren tractor was entirely different in concept to the Fowler unit. The 4-cylinder diesel engine was mounted transversely across the centre of the tractor. The cable winding drum was positioned vertically at the rear on the same plane as the engine. This imaginative layout represented a break from tradition but was indicative of sound engineering principals. It enabled the winding drum to be driven by a chain direct from the engine crankshaft.

The 1930 McLaren tractor was powered by the firm's own 4-cylinder direct injection diesel engine which developed 70 brake horsepower at 800 r.p.m. The pilot starting engine was a 2-cylinder 8 h.p. water cooled petrol unit complete with clutch and reduction gearing. It was coupled to the main engine by a Reynolds chain. Both engines were cooled from the diesel engine radiator. The gears mounted on the bracket at the front of the fuel tank were alternative change gears for the winding drum.
(From an original McLaren catalogue, courtesy W.L.Johnson)

The left hand view of the McLaren shows the location of the vertical winding drum and the enclosed chain drive from the engine crankshaft. (From an original McLaren catalogue, courtesy W.L. Johnson)

The vertical drum was supported on a shaft at either end (unlike the traditional horizontal underbelly drum) resulting in greater rigidity. The coiling of the cable was not adversely effected when the tractor had to be positioned on uneven ground and the irksome problem of the cable dropping off the drum was completely overcome. The cable was passed under the tractor from the winding drum and fed around a sheave, thence to the implement.

The McLaren 4-cylinder diesel engine had 3 cams for each exhaust valve. The cam shaft could be moved laterally, thus providing nil, half and full compression. During starting, the 2-cylinder pilot engine rotated the main engine crankshaft with the cam in the nil compression position. At 170 r.p.m. the cam was moved to the full compression position and the engine fired into life. (From an original technical manual, courtesy W.L. Johnson)

Another common factor shared between the Fowler and McLaren cable ploughing engines was that their existence within the evolution time clock of farm mechanisation was short-lived. By the mid-1930s lightweight agricultural tractors could plough an acre for a fraction of the cost of the heavy cable machines. And Harry Ferguson was about to emerge from the wings and revolutionise tractor cultivation with his hydraulically controlled 3-point linkage.

It is an inexplicable fact that few tractor publications give even scant reference to the era of the cable ploughing engines. This is regrettable, for despite their brief appearance they are an important element in the evolution of the farm tractor.

has again been restructured and Landini produces certain models painted and badged in the Massey Ferguson livery.

Landini today — a paradox

Today the Landini factory uniquely represents a strange but desirable blend of the best of the old and the best of the new production techniques. The factory is still contained within the boundaries of the quaintly old-world village of Fabbrico. Surprisingly perhaps, Fabbrico is not located adjacent to heavy industrial centres, as is customary for large tractor plants. Rather, it is deep in the countryside surrounded by well tended farms and orchards with their characteristic Italian red-roofed farm houses and patios ablaze with geraniums.

Quite extraordinarily, the factory closes for lunch! The workforce, drawn from the families of the surrounding farms, walk, bicycle or board their raucous Vespas, Ducatis or Fiats to head home for a plate of steaming pasta washed down by a quaff of vintage red, followed by a quiet siesta.

This antiquated and (most would say) unprofessional custom of closing down the plant at lunchtime is suggestive of lost productivity. To the contrary, the Landini factory pulsates with a disposition that is the envy of other tractor plants around the world. Most of the employees are the second, third or fourth generation of their family to work in the plant. They each regard it as 'My factory' and toil with a cheerful enthusiasm that is an inspiration to behold. This attitude results in a standard of workmanship and quality control of the highest order. Built into each Landini is an intangible component that is not listed in the spare parts book — *the pride of every man and woman at Fabbrico.*

The author on his farm workhorse, an 86 h.p. Landini 8860. (Photo: M. Daw)

CHAPTER 27

The English 'Caterpillar'

In the year 1932 John Fowler & Co. (Leeds) Ltd. was facing a gloomy future. The effects of the Depression were still having disastrous repercussions with both home and export sales. Steam engines, upon which the company had prospered, belonged in the past. Cable ploughing was in its death throws and Gyrotiller sales were *costing* instead of *providing*. The firm was in a loss situation and its overdraft at the Westminster Bank had to be increased, enabling trade debtors to be appeased and to provide carry-on liquidity. It was noted that Clayton Shuttleworth had discontinued production of crawler tractors and, apart from some experimentation by Ransomes & Rapier of Ipswich, no British manufacturer was producing a crawler tractor suitable for large-area farms. Accordingly, the decision was taken by the Fowler board to proceed immediately with the design of a new generation of Fowler crawlers.

The Fowler dilemma

In 1932 a Caterpillar 25 was acquired covertly by Fowler through its subsidiary, Cable Cultivating Contractors Ltd. It arrived at Long Sutton in a large crate and was duly unpacked, the exhaust pipe fitted, fuelled, and put to work. (Note: This information was obtained by the author first hand from Mr Bill Johnson, who was present on the occasion.) The Caterpillar's performance was carefully monitored, after which the unit was dismantled then scrutinised by design engineers back at the Leeds works.

The following year a prototype 25 h.p. Fowler was pronounced ready to go into production. (It had more than a passing resemblance to the Caterpillar 25!) Its diesel engine was designed by Harry Cooper, a respected senior Fowler engineer, but in this instance the unit proved unsatisfactory. The diesel injection equipment could not be calibrated accurately in relation to the air volume intake, resulting in inefficient combustion. Black smoke belching from the exhaust stack was not a sales inducement to prospective buyers. Further, poor air filtration created rapid wear in the pistons and liners. It was also found that the track gear suffered excessive deterioration due to the ingress of abrasive soils. (All this proved that the Fowler 25 was *not* a Caterpillar!)

Ironically, the Fowler 25 was well received by farmers at the 1934 Ipswich Show, although this is not surprising because the crawler certainly looked the part and it was only on static display. Had there been a field demonstration the perception would have been different.

Harry Cooper had a plan in place to rectify the engine problems when he was killed in a motor accident immediately following the Ipswich Show. This prompted the Fowler board to halt the production of crawlers, with only 30 units having been assembled.

It was not until 1936 that Fowler finally got its act together with the crawlers. Freeman Sanders, who was eventually appointed to the position of Chief Engineer, had undertaken the rectification of the problems relating to the Cooper engine and transformed it into an efficient power unit.

Two new models were introduced in 1936 which were to prove reliable and devoid of the earlier problems.

The Fowler 3-30

The medium-weight crawler was powered by a new Freeman Sanders-designed Fowler 3-cylinder 266 cu. inch diesel engine with a 4.25 x 6.25 inch bore and stroke. It was capable of delivering 34 belt h.p. at 1100 r.p.m. Weighing 9400 lbs, it achieved a drawbar pull of 6450 lbs at 1.81 m.p.h. (Note: Freeman Sanders was to later become associated with Turner and Lloyd tractors.)

A head-on view of the 1936 Fowler 3-30 with 11 inch track plates. (From original sales brochure, courtesy W. Johnson)

The Fowler 3-30 with 24 inch grouser plates. (From original sales brochure, courtesy W. Johnson)

The Fowler 4-40

The Freeman Sanders-designed engine in the Fowler 4-40 crawler was identical to that in the 3-30, except that it had an additional cylinder. The four cylinders gave a displacement of 355 cu. inches and provided 47 belt h.p. at 1100 r.p.m. The drawbar pull was tested as being 8700 lbs at 1.81 m.p.h. (This was an impressive figure when related to the horsepower.) The unit weighed 10 100 lbs when fitted with the standard width track plates of 14 inches.

The attractive modern lines of the 1936 Fowler 4-40. The Caterpillar influence in its design is clearly evident. (From an original catalogue, courtesy W. Johnson)

The Fowler 4-40 was unquestionably the best crawler produced by Fowler in the 1930s. Had Fowler concentrated on this one model instead of extending the range from 35 to 80 h.p., and undertaken a programme of establishing a competent worldwide dealer network, the firm's financial outcome could have been quite different.

CHAPTER 28

The Gregarious Garner

The four-wheeled Garner is remembered as a small tractor of considerable merit. Yet it was launched at a time when the reputation of the British tractor industry was being tarnished by an assault from a handful of questionably designed lightweight tractors. The Garner stood apart from these ill conceived counterparts as a basic but well engineered machine.

Mixed fortunes

The Garner roots extend back to 1907, when Henry Garner established a motor dealership for the purpose of selling Humber and Argyll motor cars. (The Argyll was the only car ever produced in Scotland and featured an incredibly silent sleeve valve engine of considerable sophistication.) The company prospered and in 1908 the name of Henry Garner Ltd. was registered. In that year new premises were acquired in Birmingham.

The first Garner association with tractors commenced in 1918 when a decision was made to import from the U.S.A. Galloway tractors manufactured by William Galloway & Co. of Waterloo, Iowa (see Chapter 20.) The tractors were sold in Britain with Garner livery, however the American machines were largely ignored by British farmers owing to their relatively high price. In an endeavour to encourage sales, a Galloway/Garner, amidst a great deal of fanfare, towed a mobile threshing mill non-stop from London to Birmingham. It was hoped to be able to cover the distance in one day. In fact the trip took one day and four minutes — a remarkable achievement for 1918! Despite the resulting publicity, sales did not improve. Garner persevered with the tractor until a judgment was taken in 1920 to discontinue its importation.

In addition to its involvement with motor cars, Garner also established a commercial vehicle manufacturing division in 1927. The uncertainty of fluctuating sales necessitated an injection of capital, which resulted in the company forming an alliance with the manufacturers of Sentinel steam trucks. The effects of the Depression years, however, took their toll and Garner experienced terminal fiscal problems in the 1930s which led to the liquidation of the firm in 1937.

Despite all these associated problems the reputation of Garner products remained in high esteem. It is not surprising, therefore, that a group of financiers purchased the name and established a new factory in London, enabling Garner vehicles to once again enter production. This occurred in 1937, only months after the final collapse of the original regime.

The post-war tractors

At the conclusion of its wartime military vehicle production, a subsidiary company registered as Garner Mobile Equipment Ltd. was created in 1947 for the purpose of manufacturing and distributing a new range of walk-behind 2-wheeled market garden tractors. These were introduced in 1948 and quickly found favour with commercial gardeners and nurserymen. During the first twelve months of manufacture 500 units were produced. Encouraged by the success of the 2-wheeled machines, the designers returned to their drawing boards and created a 4 wheeled stretched version. The first 4-wheeled Garner tractors were released in 1949.

Their technical specifications had a considerable degree of commonality with the earlier 2-wheeled versions. The 4-wheeled light tractor was equipped with a J.A.P. 6 h.p. air-cooled single-cylinder engine. Power to the 3 forward and 1 reverse speed gearbox was delivered through an enclosed centrifugal clutch. There was no manual control of the clutch, its action being determined by the r.p.m. of the motor. The drive was delivered from the differential to the rear wheels by roller chains. The gear shift was located to the rear of the

Stephen Allan's restored 1950 Garner is one of the few remaining. Note the mid-mounted tool bar to which a range of implements was designed to be attached.
(Photo courtesy T.O.M.M. Magazine)

CHAPTER 30

Lesser-known American Tractors

The following is a selection of particular models of classic tractors not widely known outside Canada and the U.S.A. In some instances the *make* might be familiar but the *model* relatively unknown.

1947 Empire 88

The Empire 88 was first introduced in 1946 by the Empire Tractor Corp. of Philadelphia, Pennsylvania. It was aimed at the small farm market in competition with a range of tractors that included the Ford 9N, Farmall A, Massey Harris Pony and Case S.

The unit was built along similar lines to the CO-OP, Friday and Graham Bradley tractors, utilising channel chassis construction as distinct from the unit construction adopted by the larger manufacturers. The engine and transmission of the Empire were mounted between the channels. The operator sat high on a pressed steel sprung seat. This gave excellent vision but was a handicap during orchard applications. The design of the front axle was basic but somewhat unusual, however its relatively high-clearance enabled the tractor to enter fields of semi-advanced crops without causing damage.

A 40 h.p. 4-cylinder side valve Willys Overland Jeep engine provided the power. The tractor was also fitted with a Jeep gearbox and transfer case with its 6 forward and 2 reverse speeds. This was a shrewd design initiative, as spare parts were readily available and the Jeep components were held in high regard following the vehicle's legendary performance in the war just ended.

In 1948 the Model 88 was replaced by the 90. Apart from a slight cosmetic change the only difference in the two models was lower gearing in the 90. The competition from the traditional tractor companies, however, proved too great for the small Philadelphia firm. The Empire Tractor Corp. went out of business in early 1949. This was a shame, as the tractor had considerable potential and with further development might have obtained sufficient sales to assure the financial viability of the company.

The 1947 Empire 88 pictured is owned by Mr Keith Miller, a well known identity of the city of New Sharon, Iowa. His well presented tractor is one of 86 registered by the Empire Tractor Club of Kayusa, New York State. Keith Miller's 88 is frequently displayed at county fairs, where it attracts a great deal of interest.

1918 Emerson Brantingham Model Q

In 1912 the long established farm machinery manufacturing firm of Emerson Brantingham Implement Co., Rockford, Illinois, made successful takeover bids for the Gas Traction Company of Minnesota and Reeves and Company of Indiana. Reeves had just developed a 40-65 tractor and Gas Traction was the manufacturer of the well regarded Big 4 prairie tractor. So, with the stroke of a pen, Emerson Brantingham had acquired the manufacturing rights of two tractors without having to become involved in their lengthy and costly developmental work. The firm was to continue to produce the two heavyweight tractors until 1920, although the engine for the Reeves 40-65 was manufactured by the Minneapolis Steel and Machinery Co.

Emerson Brantingham first designed and introduced its own tractor in 1916. This was the Model L, which was a medium-weight 20 h.p. tractor with a single driving rear wheel. Although it was reasonably well received by some farmers, others were discouraged by its three-wheel configuration. (Note: An Emerson Brantingham Model L in original working order is owned by a New Zealand collector.)

Accordingly, in 1917 an improved four-wheel version was released named the Model Q. This and its future variant (the Model AA) proved to be the

The Empire 88 was never tested at Nebraska, which was a detrimental factor in the minds of some prospective buyers. Despite being an attractive and functional unit it could not successfully compete against the might of Ford, Ferguson, etc. Like so many other tractors made only in small numbers, the Empire is today regarded as an extremely desirable collectable artifact.(Photo courtesy K.Miller)

LESSER-KNOWN AMERICAN TRACTORS 207

firm's best selling tractor and remained in production until Emerson Brantingham was bought out by the J. I. Case Co. of Racine, Wisconsin in 1928.

The model Q was a fine tractor, disadvantaged by its open gear final drives. (The AA featured enclosed final drives.) The two rear wheels had an uncommonly narrow track, but to compensate for stability the front axle was extra wide. The reason for the narrow rear width was to permit both back wheels to *remain up on the land* during ploughing and the front right wheel to *run in the furrow.* This kept the tractor level and caused less fatigue for the operator, but more importantly the engine oil splash feed *could not lubricate the engine if the tractor was constantly tilted.*

Dan Ehlerding of Jamestown, Ohio, is the owner of this fine example of a 1918 Emerson Brantingham Model Q 12-20 powered by the E.B. 20 h.p. 4-cylinder petrol/kero engine. The tractor features a 2-speed gearbox and a cone clutch. (Photo: I.M.J.)

A.H. McDonald and Co. of Richmond, Victoria, imported the improved Q — the Model AA — into Australia in 1923, where it was sold as the McDonald Imperial E.B. 12-20. It appears that the tractor imported retained the narrow rear wheel track but had a narrow front axle to match. Its overall width therefore was a somewhat delicately balanced 54 inches! The unit pictured is owned by Brian Bunker of Kendenup, Western Australia. (Photo courtesy B. Bunker.)

1936 Silver King R44

Unlike the majority of American tractors, the Silver King was conceived purely out of economic necessity.

The Fate Root Heath Company was located out in the wide plains of Northern Ohio in a small town named Plymouth. The manufacturing firm was the main economic reason for the town's existence. When demand for its Plymouth rail yard locomotives stalled in the Depression years the firm's President, Charles Heath, knew that if his company went under, so did the town!

Being a relatively small organisation and therefore not hampered by the tedium of a complex management structure, Fate Root Heath could and did make quick and bold decisions. A commitment was made, aimed at saving the company *and the town,* to diversify into tractor production. After all, Plymouth was surrounded by farmers.

Following a hasty restructuring of the factory, during which time the backroom boffins had been working around the clock with the design of a new tractor, the first Plymouth tractor was announced in 1932. It was powered by a Hercules IX B 4-cylinder side valve petrol engine coupled to a 4-speed transmission with a road gear giving a handy 25 m.p.h. The stylish general purpose lightweight tractor was painted silver, for no other reason than the fact that the firm's stores department had an oversupply of silver paint which they were desperate to use up. So — silver the tractors were, but with red wheels.

A 1936 Silver King R44 owned by Trevor and Jody Payne of Young, N.S.W. In that year Fate Root Heath production averaged five tractors each day. A John Deere executive visiting the plant in 1937 claimed the Silver King to be '....the best made, poorest sold tractor'. Harry Ferguson also took time to visit the Plymouth factory and was apparently impressed by the design of the lightweight machine. It is believed there are only three complete and running Silver King tractors in Australia and at least one in New Zealand. (Photo: I.M.J., with thanks to T. & J. Payne who kindly provided Silver King historical information.)

The Custom was the sort of tractor that appealed to experienced tractormen. It is still highly regarded today by those farmers who knew and drove the tractor in the 1950s. They refer to its silky smoothness and ease of driving. No official performance figures are available as the tractor was never offered for testing at Nebraska.

There were apparently some licensing arrangements in place with the Custom, as the almost identical Jumbo and Simpson tractors were produced in California by the Jumbo Steel Products Co. of Azusa.

1960 Oliver 1800

In 1960 the Oliver Corporation of Charles City, Iowa, released the 1800 as a replacement of the previous 880 series. It was available in diesel, petrol or L.P. gas versions. Additionally, it could be purchased configured specifically for wheatland, ricefield, row crop or 4-wheel-drive applications. The row crop variant had a choice of rear axles with tyre tread adjustments ranging from 65 to 100 or 80 to 120 inches. The front axle could be full width (adjustable) or tricycle.

The row crop tricycle version was unfamiliar outside North America. The majority of the 1800 series exported were diesel-powered Wheatland Models. Unlike the larger 1900 (which was equipped with a General Motors 4-53 diesel engine) the 1800 variants were all powered by 6-cylinder Oliver engines. The 265 cu. inch 6-cylinder petrol engine had an almost unbelievable thirst for fuel. It recorded the highest horsepower hours per gallon consumption ever recorded by a petrol tractor engine at the Nebraska Test Laboratory. Under full load this represented over five gallons per hour, which is considerable for a 74 p.t.o. horsepower tractor.

The 1800 weighed 8410 lbs, but in order to reduce wheel slippage whilst undergoing drawbar pull tests at Nebraska an additional 3925 lbs of ballast was added to the rear wheels. This resulted in a pull of 4427 lbs at 5.22 m.p.h. and a 4.33 % wheel slip.

The 1960 Oliver 1800 pictured was 'discovered' in 1996 parked at the side of a main road leading south from St. Joseph, Michigan. All attempts by the author to locate the owner were unsuccessful. (Photo: I.M.J.)

1930 Massey Harris General Purpose 4 W.D.

The first serious attempt by one of the big established American tractor companies to produce a 4-wheel-drive tractor was in 1930, when Massey Harris announced the General Purpose. Significantly, this was the first new model designed from the ground up by Massey Harris engineers since the firm acquired the rights to the Wallis tractor from the J.I. Case Plow Works Co. in 1928.

The General Purpose had four equal steel wheels. (Pneumatic tyres were not an option until 1936.) The 4-cylinder Hercules engine was mounted forward of the front axle. This placed around 70% of the weight on the front wheels and rendered steering a physically arduous task. In order to compensate, lever-controlled turning brakes operating on either side two wheels enabled the tractor to be slewed in a manner similar to a crawler. The grip of the lugs, however, put considerable strain on axles, clutch and brakes during slew turning.

The 1930 Massey Harris General Purpose 4-wheel-drive pictured is owned by classic tractor collector Mr C.W. Schilling of Fountaintown, Indiana, whose knowledge of classic American tractors is profound. (Photo kindly supplied by Mr C.W. Schilling)

Surprisingly, the General Purpose was described as a high-clearance unit and promoted as a row crop tractor. Perhaps if its virtues as an alternative to a crawler had been highlighted, particularly in steep or difficult country, farmers might have taken more interest. Even though its drawbar pull of 3247 lbs at 2.3 m.p.h. was a reasonable result from its 24 h.p. engine, the majority of potential buyers ended up with either a conventional two-wheel drive or a crawler.

Another sales deterrent was the fact that to proceed with the General Purpose along a sealed road meant that *four* road bands had to be fitted instead of the customary two, which most farmers considered were more than enough!

The rear view of the Massey Harris General Purpose shows the high-clearance axles, the advantage of which was negated by the low under belly clearance. (Photo: Mr C.W. Schilling)

In an endeavour to elevate sales, the General Purpose was superseded in 1936 by an upgraded model mounted on pneumatic tyres. The new machine was named the Massey Harris Four Wheel Drive. In fact, sales remained sluggish and the tractor was discontinued the following year.

The history of the development of the farm tractor contains many similar stories of basically excellent innovations that did not succeed because they were ahead of their time and the technology was not quite there to perfect the ideas.

1916 Case 10-20

When J.I.Case Company introduced the 10-20 in 1915 it proved to be a retrograde step.

The configuration of a 3-wheeled tractor with an offset single wheel out front, in line with the furrow rear wheel driver, and the third wheel being of the same diameter as the driver but narrower and non-constant driving, had been tried and abandoned by several opposition companies. Included among these were Farmer Boy, Steel King, Wolfe and Bull. The Bull certainly attracted volume sales, but mainly as the result of price and vigorous promotion — certainly not because of performance. Nevertheless, the concept of the Case 10-20 was a direct result of the Bull sales.

A feature which partly redeemed the engineering of the 10-20 was its 4-cylinder engine. It was the first lightweight 'modern' overhead valve engine designed by Case and set the pattern for the long stroke reliable Case engines for the next half century. Another redeeming feature of the 3-wheeled tractor was a lever-operated dog clutch which could engage the drive to the *land wheel* if boggy conditions were encountered. The operator's manual, however, stressed that, as there was no differential, the land wheel drive was to be engaged only occasionally and for a short duration, and whilst travelling in a straight line.

Photos A and B show both the left and right hand views respectively of the 1916 Case 10-20. The left hand view (Photo A) shows the rear land wheel which was normally a non-driver, except when the dog clutch was engaged. Notice the location of the radiator. The right hand view (Photo B) shows the operator's position and the main driving wheel. The open ring gear final drive is also visible. The unit pictured is owned by Mr C.W. Schilling of Fountaintown, Indiana. (Photos courtesy Mr C.W. Schilling)

Final drive to the rear wheel was by an exposed ring gear (common in that era). It was considered that only one forward and one reverse gear were necessary for the 10-20.

The Case 10-20 was tested at Nebraska in 1920. The 3-wheeled 20 h.p. tractor, weighing 5080 lbs in operational trim, managed a 2613 lbs drawbar pull at 2.18 m.p.h. A total of 6679 units were manufactured between 1915 and 1918.

1919 Massey Harris MH 1

Massey Harris, the giant Canadian farm machinery manufacturer, entered the tractor arena in 1917 when it commenced distributing Bull tractors in Canada. The arrangement with the Bull Tractor Company of Minneapolis turned out to be plagued with production problems and lasted just over one year.

In 1918 Massey Harris signed an agreement with the Parrett Tractor Company of Chicago to assemble the Parrett 15-25 tractor at their Toronto plant. The tractor was rebadged and sold as the MH 1. The unit had a distinctive appearance due to its side-facing front radiator and spindly 46 inch diameter front wheels.

Designed by Dent Parrett (who also designed the first of the CO-OP tractors in later years), it was powered by a transverse mounted 4-cylinder Buda petrol engine with a 4.25 x 5.5 inch bore and stroke, developing 25 belt h.p. at 1000 r.p.m. The drive was delivered to the rear wheels by a pinion gear driving an exposed ring gear. (Later production units were equipped with the final drives enclosed within a casing, giving protection against grit and mud.) A two forward and single reverse gearbox was provided.

In order to partly overcome the fuel consumption problem, Case introduced the LAH version which was powered by a Hesselman engine claimed to run on diesel fuel. The Hesselman engine was not, in fact, a compression ignition diesel engine and still relied upon spark ignition. Farmers were wary of the Hesselman principal and the tractor did not attract volume sales. Additionally, an L.P. gas-fuelled engine became available in 1952. This proved even less popular than the LAH. In total, between 1940 and 1952 around 42 000 of the LA Series, including all engine variations, were built.

(Note: The firm of Scoble and Nash Pty. Ltd., of Parkes, New South Wales specialised in the conversion of production petrol/kero LA tractors to diesel power by installing 100 h.p. Perkins 6-354 engines. Increasing the load by 100% onto the transmission drives seemed to cause no problems — testimony to the ruggedness of the Case-designed transmission.)

By 1952 farmers were turning away from large petrol/kero tractors and directing their interest towards diesel-powered machines. It was necessary for Case to cater to this new demand and, in 1953, the company released the 500, which was essentially an LA powered by an all-new Case 6-cylinder diesel engine.

The Case 500 pictured has been fastidiously rebuilt by Robbie Morgan of Echunga, South Australia, a young man who devotes all his spare time to the preservation of classic tractors. (Photo courtesy R. Morgan)

The new engine was an outstanding example of the latest in diesel engine technology. The six cylinders were cast in pairs with a 4 x 5 inch bore and stroke and a total capacity of 377 cu. inches. It was governed to a relaxed 1350 r.p.m. and developed 64.81 belt h.p. Gear speeds produced 5 m.p.h. in third and 10.19 m.p.h. in fourth gear, an improvement over the LA. The final drive chain transmission which had been so successful in the LA and the earlier L was retained in the 500.

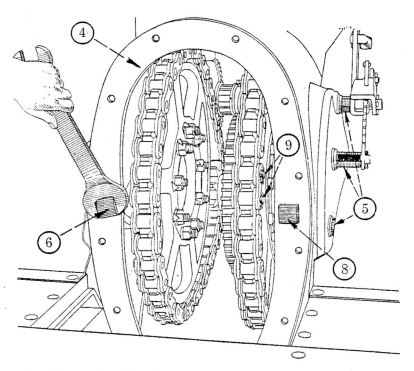

The drive chains of the Case L, LA, and 500, showing their adjustment. (Taken from Model L instruction manual)

Claimed by Case to be a first, the 500 featured hydraulically assisted power steering. This was certainly a most appreciated addition to the big tractor, which had an operational weight of 11 605 lbs.

Specifications alone fail to present a true indication of the ability of the Case 500. Its drawbar pull of 7409 lbs at 2.2 m.p.h. was impressive, exceeding that of the LA by around 1000 lbs, but in practical field conditions it was actually *half as powerful again* owing to the lugging torque characteristics of the diesel engine, which peaked at 411.3 foot pounds at a mere 1048 r.p.m.

Only a small percentage of Case 500 tractors were exported (thus their eligibility for this chapter) and they are not commonly seen in collections outside North America. The Model LA, on the other hand, is relatively commonplace although still classed as highly collectable.

1919 Illinois Super Drive 18-30

The Illinois Silo Co. of Bloomington, Illinois, entered the tractor scene in 1916 with a diminutive and short-lived 8 h.p. motorised cultivator. An improved model followed in 1917 and in 1918 the firm (now the Illinois Silo and Tractor Co.) released a well designed 30 h.p. conventional tractor. The following year, 1919, the Super Drive 18-30 was introduced. The company had again changed its name, this time to the Illinois Tractor Co.

The Super Drive 18-30 was fitted with a 4-cylinder Climax engine which had a 5 x 6.5 inch bore and stroke. (This engine was also fitted to both the Fitch and Square Turn tractors.) The two forward gear speeds could have their ratios altered by swapping two gear wheels mounted externally on the left side of the transmission housing.

The 1919 Illinois Super Drive 18-30, although not widely known, was one of the better tractors marketed in the U.S.A. in 1919. Its transmission was particularly robust and could have accommodated a much larger engine than the 30 h.p. Climax. (Sketch by Brad Drury, Taree, N.S.W.)

The 18-30 dispensed with the ring gear exposed final drives of the previous model and instead incorporated the latest internal planetary gear design mounted on each rear axle (similar to the Fiat 703 of the same period). The rear wheels were attached by an ingenious spider coupling which was spring loaded and absorbed some of the jarring normally transmitted to the operator.

Australian farmers were to know the Super Drive in a different guise. Ronaldson Bros. & Tippett of Ballarat, Victoria, entered into an agreement with the Illinois Tractor Co. to import Super Drive tractors into Australia — but fitted with Wisconsin engines. This stipulation was because Ronaldson Tippett was the Australian importer of Wisconsin engines.

The first of the Super Drives arrived at Ballarat in 1924. They were distinctive in appearance from the American models by having a larger capacity cast radiator header tank. This was designed specifically for the hot Australian summer conditions. Also, the shock absorbing rear wheel spiders were omitted — suggesting that such *sissy* additions were not necessary for the *rugged* Aussie tractormen. The imported tractor was known as the Ronaldson Tippett Super Drive 18-30 and sold from Ballarat until 1937. By that year the tractor was completely manufactured at Ballarat, apart from the Wisconsin engine and a few imported parts. (See Chapter 18.)

A rare sight. An Australian Ronaldson Tippett Super Drive in Northern Ireland! Perhaps the only example in the northern hemisphere, the Super Drive is owned by Ron and Margaret Deering and displayed at their Ballycastle House Museum, overlooking the beautiful Strangford Lough, 14 miles south of Belfast. The museum houses a fascinating collection of early tractors, including an H.S.C.S. Note: Ballycastle was part of the estate of Lord Castlereagh, the noted explorer who 'discovered' much of western New South Wales. The area still contains many Australian native eucalyptus trees grown from seeds taken home by Lord Castlereagh. (Photo courtesy R. Deering)

1925 (?) Holt 2 TON

Owen Triggell, acknowledged as the Caterpillar tractor serial number expert in Australia, has traced a total of 111 remaining complete or remnant Holt and Caterpillar type 2 TON tractors in this country. Therefore, whilst by no means rare, the figure indicates that only a small percentage of the nation's classic tractor collectors can boast a 2 TON in their shed. Then if one disregards the number of remnants or wrecks and considers only complete and running examples, the figure would likely be around 10 or 20 units.

Mr David Miller of Carisbrook, Victoria, owns what appears to be a 1925 vintage Holt Caterpillar 2 TON. There is some doubt as to its precise age as, regrettably, the vital identification numbers are missing. To further complicate the issue, 1925 is the year that the financially troubled Holt Manufacturing Co. of Peoria, Illinois, merged with the C.L. Best Gas Traction Co. of San Leandro, California to form the Caterpillar Tractor Co. The Holt 2 TON Caterpillar (Caterpillar was a Holt registered trade mark) was continued in production until 1928. A unit submitted for testing at Nebraska in 1925 was listed under Test No. 86 as being a Caterpillar 2 TON manufactured by the Holt Manufacturing Co. of Peoria.

The Holt Caterpillar 2 TON takes pride of place in David Miller's collection. (Photo courtesy D. Miller)

The slight confusion over the tractor's birth date does not detract from the fact that David Miller's 2 TON is a most desirable tractor to have in a collection. The unit is indicative of the advanced stage crawler designs had reached in only around 25 years of serious development.

The tractor was fitted with a 4-cylinder engine with a 4 x 5.5. inch bore and stroke rated at 25-38 h.p. at 1000 r.p.m. It was fitted with a 3- speed gearbox and achieved a drawbar pull of 3275 lbs at 1.77 m.p.h., whilst weighing 4040 lbs.

The 2 TON did much to promote the versatility of the crawler tractor in agricultural applications. Its horsepower, size and drawbar pull were just right for most farmers. It should be remembered that in 1925 pneumatic tyres had not yet been considered for agricultural tractors. Therefore the choice was limited to either steel lugged wheels or crawler tracks. In practice it was easier for a salesman to influence a farmer towards crawler tracks against the alternative of steel wheels than it was in later days when pneumatics, with all their compelling advantages, became available. Crawler tracks did not *pick up* clay soil in the same manner as steel lugged wheels. In addition, the weight per square inch factor was substantially reduced as there was a greater area in contact with the ground than in the case of wheels. Many farmers saw this as an advantage owing to the reduced degree of soil compaction. It is also true to say that the operator enjoyed a less jarring ride on a track machine than upon its steel-wheeled alternative.

1915 Common Sense V8

The mention of a tractor with a V8 engine (8 cylinders vee configuration) conjures up visions of a modern New Holland Versatile or a Belarus 7000. Few imagine that the first production V8-engined tractor emerged in 1915. This was in a period when tractor companies were wrestling with the complexities of 4-cylinder engines and indeed others chose to remain with the more basic single- or twin-cylinder power units.

Harry Wilbur Adams obtained his experience in tractors as a partner in the firm of Adams-Farnham Co. of Minneapolis. The firm, founded in 1909, specialised in the production of steam engine components and complete petrol engines. Its venture into heavyweight petrol-engined farm tractors in 1913 proved an expensive exercise and resulted in the company's closure shortly after.

H.W. Adams took his expertise to the Common Sense Gas Tractor Co., also of Minneapolis, and contributed to the development of the Common Sense Tractor, powered by the firm's own side valve V8 engine. The remarkably advanced engine had a 3.25 x 5 inch bore and stroke and produced an impressive 40 belt h.p. at 1200 r.p.m. By over-riding the governor and increasing the r.p.m. the engine was capable of delivering a whopping 70 b.h.p. This was the sort of performance that hitherto could only have been obtained from massive engines such as those used by Twin City, Holt, Aultman Taylor, Best, etc., weighing around six times that of the V8.

The Common Sense had a single centrally located rear wheel with an exposed chain drive from the 2-forward-speed gearbox. Although no data is available it is obvious that the 3-ton tractor must have been an impressively aggressive performer, more akin to modern tractors.

Ironically, it was the advanced technology of the engine that was responsible for the tractor's disappointing sales. Farmers were suspicious of, and concerned by, the intricacies of a V8 engine — that is until Adams thought up a plan to change their attitudes.

He established a tractor training school in Minneapolis in 1917, aimed at attracting farmers who really would like to have owned a tractor but were hesitant owing to their lack of experience and knowledge. Adams was swamped by eager farmers, particularly of the younger generation, from all over Minneapolis and adjacent states, all clamouring to learn about tractors.

Adams was proved a convincing lecturer and when he pointed out to his enthusiastic, attentive pupils that a tractor was no good unless it had a V8 engine, *his message was accepted as gospel.* Of course there was only one tractor available fitted with a V8 engine!

The result of Adam's teachings certainly gave a considerable acceleration to the sales of Common Sense tractors. For a while the firm flourished, however its newfound impetus was short-lived. The firm was taken over in 1919 by the Farm Power Sales Company and the production of Common Sense tractors discontinued soon after.

The 1915 Common Sense tractor from Minneapolis had sleek lines and provided the driver with excellent vision — not altogether common for 40 h.p. tractors of that period. It was also relatively uncommon to be able to mount a tractor from the side in that era. (Line drawing by Brad Drury, Taree, N.S.W.)

CHAPTER 31

Ferguson Model A in Focus

My thanks to Mr Selwyn Houghton of the Ferguson Self Help Group for permission to reproduce these historic photos from his collection of Ferguson memorabilia. Thanks also go to Mr Ronnie Deering of Ballycastle House Farm Museum, Newtownards, Northern Ireland, for making it all possible.

The Ferguson System

Designed by Harry Ferguson and manufactured by David Brown, the Ferguson A was the first tractor ever to have a 3-point linkage implement attachment system with a draught response hydraulic control. This was the famous Ferguson System that revolutionised mechanised farming throughout the world (see Chapter 19).

Released in 1936, the little grey tractor was an improved version of the original Black Tractor prototype (which is today preserved in the British Science Museum). Farmers were sceptical of the pulling ability of the Model A in relation to its diminutive weight of 16 cwt. But owing to the brilliance of its weight transfer system, which caused the implement to transfer its

The Ferguson assembly line in the David Brown factory. It is astonishing that mudguards were not considered a standard fitting on these 1936 tractors. At least an operator would have a good incentive not to fall asleep!

resistant forces onto the tractor thus adding 'weight' to the wheels, the lightweight machine could match the traction and resulting drawbar pull of tractors twice its weight whilst consuming half the fuel.

The initial 500 tractors were powered by a Coventry Climax engine and the remaining 850 were equipped with a David Brown engine of 2010 c.c. developing 20 h.p. at 1400 r.p.m. The gearbox provided 3 forward and 1 reverse speed. The retail price in Britain was £224.

Ferguson Model A tractors ready for transportation from the Huddersfield Works of David Brown & Sons Ltd. Note the Scammel Scarab 3-wheeled tug lorry coupled to its articulated trailer. With their single front wheel the Scammels could execute a turning circle within their own length. They were the perfect transport vehicle for negotiating narrow warehouse courtyards originally designed for horse traffic. All the major British railway companies deployed fleets of Scammel Scarabs. The two vehicles pictured were owned by the the London Midland Scottish Railway Company, indicating that their cargoes of tractors and implements were heading for a rail journey. The tractors equipped with steel wheels are fitted with flat steel bands covering the lugs on the rears. These bands were fitted during transportation to avoid damaging the timber decking of the lorries.

Mr Harry Ferguson demonstrating the manoeuvrability of his Model A to a group of attentive Herefordshire farmers. Note the absence of mudguards. The venue is Dormington Court Farm, Herefordshire.

A good view of the Ferguson 3-point linkage system. It is unfortunate that Harry Ferguson did not divert some of his genius into designing a more comfortable seat. The reason for the presence of the bottle jack under the left axle is unknown.

The wee lads are clearly fascinated by the way the Ferguson 7-tyne spring cultivator has penetrated the ground whilst the tractor is stationary. The ground must either have been extremely loose or Mr Ferguson committed the sin of backing the tractor with the implement in the ground. Mr Ferguson's attire of breeches, riding boots and hacking jacket was the garb of a 'gentleman farmer' in the 1930s.

The 2-furrow Ferguson mounted mouldboard plough was in itself an engineering triumph. It replaced the heavy trailing draught plough design that required pulling by a much larger tractor. The operator is turning a neat furrow — but then failure to do so would have evoked disdainful censure from the critical farmers. All Ferguson implements were manufactured from an extremely high grade lightweight steel that usually managed to resist breaking or twisting even in the hands of a rough operator. Should welding ever be required it took the skills of an experienced tradesman to work the high grade steel.

The demonstration by the young lady, groomed as if for afternoon tea at the Savoy, was aimed at proving the ease of driving a Ferguson tractor. In 1937 most of the farmers would no doubt have been amused or even shocked by such a wanton display of femininity.

Harry Ferguson was a skilled operator and took delight in personally demonstrating the advantages of the Ferguson System of farming. This 7-tyne spring cultivator was priced at £26, which was the price of all Ferguson implements in 1936. Note the build-up of sticky clay between the rear wheel lugs. A tedious job for an operator of any tractor not fitted with pneumatic tyres working in clay soil, was the regular requirement to remove the packed clay with a crowbar. This could take ages and, if ignored, the wheels would pack up with clay to such an extent that the traction and gearing ratios were significantly affected.

Harry Ferguson invited a farmer to experience the ease of ploughing with the Model A. He then held the fence post halfway across the furrow in the full knowledge that the light precise steering of the tractor would enable the farmer to drive the front furrow wheel between the post and furrow wall without touching the post.

The first Hanomag tractors

The company had experimented with tractors during the early part of the 20th century and in 1912 produced the transmission for the giant 3-wheeled Wendeler-Dohrn motorised plough. But it was not until 1924 that Hanomag seriously entered the farm tractor arena. In that year it released the 28 h.p. 4-cylinder petrol-powered Model WD plus a crawler, the Z25.

1912 Wendeler-Dohrn motorised 5-furrow plough. (Hanomag promotional material)

1924 Hanomag Z25 28 h.p. crawler. (Hanomag promotional material)

1935 Hanomag Diesel 30 h.p. Note the down swept exhaust pipe. (Hanomag promotional material)

The first Hanomag tractors arrived in Australia around 1935. The initial model released by the importers, Demco Machinery Co. Pty. Ltd., was powered by a 4-cylinder full compression ignition diesel engine producing 30 drawbar horsepower at 1300 at r.p.m.

Although the tractor was magnificently engineered, few were purchased by Australian farmers owing to the high price tag when compared to American imports. But, more significantly, it was because of the apprehension exhibited by farmers towards the complexities of diesel power.

The post-war period

The release of the R40 in 1949 saw a resumption of the importation of Hanomags into Australia following their absence during World War II. The R40 was equipped with a 5.2 litre 4-cylinder diesel engine rated at 40 h.p. and the tractor could be ordered on either steel lugged wheels or pneumatic tyres. In the case of the former, the normal 5-speed gearbox had the 4th and 5th gears eliminated, as travelling at high speeds on steel wheels would have shaken the tractor to pieces.

In 1952 Hanomag trebled its previous year's tractor production.

1949 Hanomag R40 fitted with spring leaf front axle suspension. (Hanomag promotional material)

1955 Hanomag row crop tractors could not compete on a price basis with the comparable Farmall and John Deere models. (Photo: Hanomag tractor brochure)

CHAPTER 35

Other Interesting European Classics

In 1950 there were over 50 European tractor manufacturers. The total of their model variants exceeded 400. This certainly presented a somewhat bewildering choice for some farmers, although it should be understood that many of these manufacturers were relatively small parochial firms, selling mainly into their own areas.

This chapter examines a selection of technically interesting European tractors not covered elsewhere within these pages.

Same

The Italian firm of Same S.P.A. can trace its ancestry to Giovanne and Francesco Cassani who, in 1927, introduced their first agricultural tractor. Significantly, it was equipped with a diesel engine. The Cassani firm at Treviglio was restructured in 1942 and the name changed to Same. In later years the Lamborghini Tractor Co. was acquired and in 1996 the historic firm of Deutz was also to come under the Same corporate umbrella.

For the record it should be mentioned that some Same sales brochures refer to the Cassani tractor as being the world's first *diesel-powered* tractor. This is incorrect, as the Benz Sendling diesel tractor preceded the Cassani by around four years. It is true to state, however, that the Cassani was the first full compression ignition diesel engined *Italian* tractor.

A rare early Same has recently surfaced from obscurity in South Australia. The little tractor, owned by Tony Clements, is undated, but from information kindly made available to the author by Same importer, Murray Tractor Importers Pty Ltd of Robinson, N.S.W., it seems that the South Australian tractor is a later 4-wheeled version of a 3-wheeled Same released in 1942.

Tony Clements' rare machine is powered by an air-cooled single-cylinder petrol/kero o.h.v. 4-stroke engine, equipped with a decompressor lever to aid cranking. Surprisingly, it features a 2-stage clutch with a live p.t.o. shaft. The gearbox has a high and low ratio providing six forward and two reverse speeds.

No doubt there are a few identical Same tractors tucked away in Italian barns. In Australia, however, Tony Clements' classic is an exceptional gem and almost certainly the oldest Same in the country.

The single-cylinder Same 'discovered' in South Australia and now owned by Tony Clements. (Photo: T. Clements)

The photo of Peter Murray of Robinson, N.S.W. standing alongside a Same 190 Titan graphically illustrates the Same evolution when compared with Tony Clements' tractor. (Photo: I.M.J.)

Steyr Daimler

The Austrian organisation of Steyr Daimler Puch can chart its beginnings back to 1894. By the 1930s it had developed into a major manufacturing conglomerate involved in every facet of automotive transportation.

Moffat Virtue Ltd. first introduced Steyr tractors into Australia in 1952. This was during a period of fierce competition between Ferguson, David Brown, International Harvester and Fordson, all contesting vigorously for the small-tractor segment of a very lucrative market. The larger of the two imported Steyr models, the 180, fell into direct competition with those mentioned. The smaller 80, however, with only 13 belt h.p., belonged to that somewhat vague category of being too small for *real* work and too expensive for the average market gardener.

There is no arguing that the Steyr Daimler tractors were quality all through. The reasons they did not make much of an impression in the Australian

OTHER INTERESTING EUROPEAN CLASSICS

market was a combination of factors, including the unfamiliar name of Steyr Daimler, the competition from the big-name and less expensive opposition brands, but principally on account of Moffat Virtue Ltd. Although an old and respected company, Moffat Virtue was more attuned to marketing shearing gear, pumps, stationary engines etc., and failed to grasp the complexities involved in the promotion and field back-up servicing of tractors.

1954 Steyr Daimler 80, powered by a liquid-cooled 1330 c.c. single-cylinder, 13 belt h.p., diesel engine, with a 4-forward-speed gearbox, 3 p.l. and p.t.o. Its drawbar pull was 1872 lbs at 1.8 m.p.h. The tractor has been restored and is owned by Colin Roberts of Nambour, Queensland. (Photo: C. Roberts)

1955 Steyr Daimler 180 powered by a liquid-cooled 2666 c.c. twin-cylinder 26 belt h.p. diesel engine, with 5-forward-speed gearbox, 3 p.l. and p.t.o. The tractor has been restored and is owned by Trevor and Jodi Payne of Young, N.S.W. (Photo: I.M.J.)

Benz Sendling

The introduction of a 3-wheeled tractor in 1923 by the Berlin firm of Benz Sendling made history. It was claimed to be the world's first tractor to enter production powered by a full compression ignition diesel engine.

In fact, the Benz Sendling was an ungainly difficult-to-drive monstrosity. It was unstable to the degree that it required foldout outrigger jockey wheels to prevent the machine from capsizing on anything but flat ground. As a consequence, from 1925 the 3-wheel configuration was changed to a more conventional 4-wheel layout.

The 2-cylinder 4-stroke 5.73 litre diesel engine produced 32 h.p. at 800 r.p.m. It was always cantankerous to start and required the assistance of a sparking ignition paper inserted into the head of each cylinder in order that combustion would take place when the engine was cold.

Australia is fortunate to have both a 3- and 4-wheel version of these very rare historic tractors.

A Benz Sendling, sold new in 1926, is but one of the priceless collection of classic tractors held in trust by the Booleroo Steam & Traction Preservation Society Inc. of South Australia. The driver of the tractor is Greg McCallum of Ororoo, a fine young man with a sincere desire to continue the family tradition of restoring and preserving the classic tractors that are an important part of South Australia's farming heritage. (Photo: I.M.J.)

Eicher

In the late 1940s British farmers were benefiting from the new technology being introduced by Ferguson, David Brown, etc. They were, however, also being subjected to a variety of poorly engineered lightweight tractors that had been rushed into the post-war market, such as the B.M.B. Brockhouse President and Newman. European farmers, on the other hand, had a wealth of excellent newly introduced tractors from which to choose. Gebr. Eicher of Forstern was an example of a German manufacturer that produced a splendid range of diesel-powered machines during that period.

Throughout Germany to this day, countless Eicher tractors continue to give consistently reliable service. Surprisingly, Australia did not have a volume importer of Eichers, possibly on account of the production output being largely absorbed by European farmers. Nevertheless, a few Eichers did in fact find their way Down Under.

A 1956 Eicher Model EAKL 1509 was privately imported by a New South Wales German immigrant farmer onto his property at Captains Flat, N.S.W. Today the tractor is carefully preserved by his son, Michael Wurzer, and is used regularly on his farm.

Mal Cameron of South West Rocks, N.S.W., recently unearthed a 1960 Eicher type EM200 30 h.p. 2-cylinder tractor in Queensland, which he was able to purchase. It is currently in its final stages of restoration — due for completion 1997. It appears that the unit was brought into Australia as a secondhand tractor. Interestingly, the 1.43-ton machine is equipped with a ZF transmission and a Bosch-designed hydraulic system and 3-point linkage. According to Mal Cameron, who is a veteran of scores of classic tractor and

1956 14 h.p. Eicher owned by Michael Wurzer. (Photo: I.M.J.)

1960 30 h.p. Eicher owned by Mal Cameron. (Photo: M. Cameron)

engine restorations, the quality of workmanship that went into the original production of his Eicher is representative of the very best of Teutonic engineering.

Indian-made Eicher tractors were exhibited to European tractor distributors at the 1984 Hanover Trade Fair. They were made under a licensing arrangement by Eicher Goodearth at its Faridabad Plant. Eicher enthusiasts, accustomed to the high quality of the Forstern-made Eicher tractors, must have been shocked and dismayed when they inspected the Indian machines. Their lack of quality control and overall finish was almost beyond belief. (Note: The author, who attended the Hanover Fair of that year, had never in his experience inspected such shoddily made machinery. Admittedly he had not then been exposed to some of the Chinese tractors that were exhibited in Australia in 1994!) It should be stated though, in defence of Eicher Goodearth, that in 1984 the Indian-made Eichers accounted for an 18% share of the local Indian market and 14 000 tractors were produced at Faridabad that year.

Holder Cultitrac

The long established German agricultural machinery manufacturer Holder GmbH of Grunbach, launched an imaginative range of lightweight 4-wheel-drive tractors in the early 1950s. Of particularly significance was the articulated (or hinged) chassis design, which had in fact been pioneered in Europe by Lanz 30 years previously.

The Model A12 (pictured) is believed to have been imported into Australia around 1958. It is powered by a single-cylinder 12 h.p. Sachs 2-stroke air-cooled diesel engine vertically mounted forward of the front axle.

Technically the Holder A12 was advanced for its time. The transmission consisted of 6 forward and 3 reverse speeds and it was provided with a 2-speed p.t.o. shaft. A differential lock, 4-wheel brakes and 3-point linkage were all listed as standard equipment. The ZF pivot joint steering gave a turning circle of a mere 5 feet.

The Holder Cultitrac A12 is symbolic of the high standard of post-war German technology. The unit featured is owned by Lance and Elenore Wilson, proprietors of the TOMM Magazine. (Photo: L. Wilson)

The 12-horsepower Sachs engine of the Holder A12 gives excellent stability and traction on level country, by being mounted forward of the front axle (Photo: L. Wilson)

The principal importers of the Holder into Australia were Stuart Fell & Co. of Melbourne, Dominion Motors (Qld.) Pty. Ltd., of Brisbane, Western Diesel Service of Rydalmere, N.S.W. and A.H. McDonald (Sales) Pty. Ltd. through their Western Australia branch. Unfortunately for the image of Holder, it was often promoted as an ideal hillside tractor (Italian articulated tractors were similarly promoted in Australia). In fact, articulated tractors in inexperienced hands can be quite dangerous on steep country due to the changing centre of balance each time the steering wheel is turned.

L.H.B.

In 1951 the German firm of Linke Hoffman Busch of Braunschweig commenced production of what was to be a run of 400 tractors. The company had excellent credentials, as it was well known for its range of agricultural implements and tractor trailers. There appears to have been an association with the firm of Kurt Kogel of Munich as the Bavarian organisation made a nearly identical tractor.

Despite being of unit frame construction, the L.H.B. Model LHS-25 was largely made up of components purchased from other manufacturers. The engine chosen was a Henschel 4-stroke liquid-cooled diesel (Henschel was widely known throughout Europe for its steam locomotives, trucks and World War II military aircraft). Its two cylinders developed 22 h.p. at 1800 r.p.m. The smoothly operating 5-forward and 1-reverse-speed gearbox was a product of the ZF company and gave a handy road speed of 14.92 m.p.h. Drawbar pull was an estimated 3,120 lbs at 1.86 m.p.h. A differential lock and turning brakes were standard equipment but a hydraulic 3 p.l. system was classed as an optional extra.

1951 L.H.B. Model LHS-25. The Henschel engine is clearly shown in the photo. (Photo: I.M.J., courtesy A. Latimore)

A bid was made to market the L.H.B. in Britain. However, the disappointing sales experiences of Allgair, Fahr, Hanomag, and Porsche did not encourage the further establishment of German tractor dealerships in the U.K. at that time. Severe British import restrictions and high tariffs put the German imports at a considerable price disadvantage compared with their British counterparts.

What is believed to be the only example of an L.H.B. in Australia is to be found in the Alan Latimore collection at Comboyne, New South Wales.

Nord A.D.N. 25

In 1951 Tractors Diesels and Equipment Pty. Ltd., of Balmain, N.S.W. imported four crawler tractor models manufactured in France by Acies Du Nord. They ranged from 30 to 130 h.p. During that period the European crawler tractor reputation (apart from Hanomag) was at an all-time low in Australia owing to the presence of Ansaldo, Breda, Continental and Vender imports. The Nord A.D.N. did nothing to enhance the perception.

The all-embracing reputation unfortunately included the smallest of the Nord range, the A.D.N. 25. This was indeed regrettable as the 25 was a commendable little crawler. Under different circumstances it might have sold strongly against the Bristol 20, John Deere MC and Oliver HG. The heart of the A.D.N. 25 was a first class 2-cylinder 4-stroke liquid-cooled diesel engine, capable of delivering an aggressive 30 h.p. In first gear at 2.4 m.p.h. it exerted an impressive 4100 lbs drawbar pull.

Some intending buyers of the little French crawler might have been dissuaded by the unusually small diameter drive sprockets and front idlers. Actually this feature had several advantages, including easy access for the operator and a low centre of gravity making, it very suitable for hillside operation.

Few Nord crawlers remain in existence today and those that do are greatly prized by their owners. Les Knoll of Toowoomba is fortunate to own the A.D.N. 25 pictured.

The miniature-sized A.D.N. 25 was the pick of the Nord range. (Photo: I.M.J., courtesy Les Knoll)

Munktells

Johan Theofron Munktell established his business in 1832 in the Swedish town of Eskilstuna. Like so many of the tractor manufacturers with origins extending back into the early 19th century, the Munktells organisation's rise to greatness was due to its involvement with the production of steam engines. Internal combustion powered Munktells tractors did not emerge until 1913 when the Eskilstuna factory embarked on a program of producing heavyweight machines similar to the Marshall Colonial and the Emerson Brantingham Big Four. The Munktells engine was a vertical hot-bulb 2-cylinder semi diesel 2-stroke of a whopping 14.4 litres capacity producing 40 h.p.

The precedent was set for the Swedish firm to continue development of this type of engine. A 32 h.p. version was used in the Munktells type 25 which first appeared in 1934. It, too, featured a 2-cylinder 2-stroke semi diesel.

In terms of sales, the type 25 was a successful tractor, with 1134 units having been produced when the series was discontinued in 1938. It was well suited to the cold northern climate, although it can be assumed that the eight units imported into Australia by Demco Machinery Co. Pty Ltd performed satisfactorily in the hot Australian inland conditions, for at least three have survived and are in good going order. (Demco also imported Hanomag and Lanz during the same period.)

Starting the engine was a complicated procedure and involved the use of a built-in blowlamp for preheating the two hot bulbs. Compressed air pressured to 260 p.s.i. was then released into the front cylinder, which impulsed the engine into firing. The rear cylinder was used as a pump to compress the air into the receiver (storage tank) prior to stopping the engine, in readiness for the next start. If the receiver had not been charged, or had the pressure escaped, it was necessary to use a hand-held blowlamp and crank the engine by hand.

The Munktells Type 25 had 4 forward and 1 reverse speed. Its lugging performance, like all tractors equipped with semi diesel 2-stroke scavenging system engines, was outstanding and considerably greater than a petrol-engined tractor of similar horsepower. The low-grade fuel oil with which it was fuelled resulted in low cost running, an important factor to farmers during the Depression years.

In 1939 the new Types BM2 and BM3 replaced the Type 25.

A 1938 Munktells Type 25, serial no. 7725, owned by prominent Queensland peanut farmers Graham and Douglas Peterson of Kingaroy, Queensland. The driver of the tractor is Doug Peterson. The tractor's 2-cylinder engine has a 150 mm bore and stroke and develops 33 h.p. at 900 r.p.m.
(Photo: G. Peterson)

CHAPTER 36

A Personal Recollection

The Christmas Tractor

The author remembers a particular Christmas and how it was irrevocably linked to an ageing tractor. (A condensed version of this story appeared in *Australian Grain*, December 1995.)

The white landscape

Each year as the Yuletide Season comes around I find myself reflecting upon the warm, nostalgic memories of the long gone Christmas days of my boyhood in Scotland. These were magic times.

December 1947 was a particularly bitter winter, with the snow arriving early. Our village, blanketed in snow with sparkling icicles hanging from the eaves of the little stone cottages, presented a scene of pious tranquillity. The surrounding hills looked stark and bleak as they froze under their mantle of sullen clouds and drifting snow. The lonely farms snuggled in their glens whilst, inside the byres, cattle huddled together for warmth.

Trout streams and burns, normally rushing and tumultuous, were frozen into a solid, eerie silence. The great sycamore and chestnut trees stood naked like giant black sentinels, silhouetted against the white landscape. Many of the farms had become isolated from the outside world as blizzards created deep drifts across low gullies along winding roads. As a consequence, it appeared that many of the families were going to miss out on receiving their Christmas mail, which of course would include large gaily wrapped parcels of toys for the children, sent by distant relatives.

'A wee bitty o' snow'

Mr Adamson was a man not easily daunted by a 'wee bitty o' snow'. For 30 years he had driven the local Post Office van and was proud of his unblemished record of delivering the Royal Mail on schedule. Accordingly, on Christmas Eve, having waited all morning in the vain hope that the weather would clear, Mr. Anderson fitted snow chains to the tyres of his bright red Morris van and headed it out of the village for the treacherous Glen Road.

Despite some anxious moments of slithering and broadsiding, the little Morris continued determinedly onwards as Mr. Adamson coaxed it gently through the driving snow. All went well until, halfway up a precipitous climb, the wheels lost their grip on a patch of black ice and the vehicle slithered sideways into a frozen ditch.

Mr Adamson considered his situation as the snow quickly obliterated his vision through the windscreen. The village was a long way back, but Kirkwood Farm was somewhere over the next rise. In fact it took the middle-aged and rotund Mr Adamson the best part of an hour's arduous trekking through waist-deep snow to reach the sanctuary of Kirkwood. He had frequently been obliged to seek shelter from the slanting snow in the lea of the dry stone walls which encompassed the fields.

'..... the little Morris continued determinedly onwards ......'

Without hesitation, upon learning of the mishap with the van, Jock Wallace, Kirkwood's tenant farmer, offered Mr Adamson the services of a gleaming grey Ferguson tractor, purchased a few months earlier in time to 'howk' the 30 acres of Golden Wonder potatoes, for which Kirkwood Farm was noted.

The Ferguson was housed in a disused stable and no time was lost in firing it into life. It was moved out into the blizzard and pointed in the direction of the Post Office van. The sweetly running little tractor had barely cleared the farm gate when it slid gracefully off the road, nearly disappearing into a deep snow drift. Mr Adamson's worthy intentions of delivering the Christmas mail appeared doomed! The situation was indeed desperate. Could he trek the additional mile up the glen to Easter Colzie and seek assistance at the MacGregor Farm?

Hamish MacGregor and his Fordson

Hamish MacGregor was an old-fashioned farmer. Never would he have considered exchanging his paraffin lamps for newfangled electric lights. Nor would he contemplate for one moment trading in his 1926 Model F Fordson tractor, complete with steel wheels equipped with dangerous-looking steel grips, for a more modern machine. Nevertheless Hamish was a kindly man and an imposing figure with his weatherbeaten, crinkled face and long white beard. So when he answered a knock on the thick oak door of his whitewashed cottage and was confronted by an exhausted and shivering Mr Adamson, he immediately responded to the urgent request for help.

Firstly, Hamish poured a liberal dram of malt whisky for himself and Mr Adamson as fortification against the bitter cold. He then donned a bright red lumber jacket with its Arctic Fox collar and led the way to his shed which contained the Fordson tractor. By now it was late afternoon and dark. With the aid of the flickering light from a paraffin storm lantern, buckets of steaming hot water were poured into the radiator of the ancient machine. After three cranks of the handle with the spark 'off', followed by two swings with the spark 'on', the engine roared into life, engulfing the men in clouds of choking fumes entrapped within the confines of the diminutive shed.

The doors were flung open and Hamish emerged from the dense smoke into the black night upon his tractor, with Mr Adamson grimly holding on as he stood on the rear drawbar. Hamish had crudely secured the storm lantern, by a length of binder twine, to the front of the radiator in an endeavour to illuminate his way to the now almost completely buried Post Office van.

'All through the latter hours of that Christmas Eve the Fordson spluttered its way to the farms in the glen.'

The steel-cleated wheels proved their worth in the snow. They obtained a positive grip and even when treacherous black ice was encountered, the tractor negotiated a straight course without hesitation. Eventually and with difficulty the abandoned van was located. Its contents were transferred and lashed somewhat precariously to the rear of the Fordson, using a hay-rick hemp rope. When Hamish finally clambered upon the cast iron seat, with Mr Adamson balancing on the drawbar, they were completely surrounded by the sacks of mail and bright packages. (Although it was not known at that time, the Morris was destined to remain buried until the thaw arrived the following March.)

All through the latter hours of that Christmas Eve the Fordson spluttered its way to the farms in the glen. Hamish somehow managed to negotiate the appalling conditions by peering into the dim pool of light in front of the tractor created by the feeble lamp.

Enchanted wide-eyed children gazed from their upstairs bedroom windows in awe. They could hardly believe their eyes as they saw a huddled white bearded figure, wrapped in his bright red jacket with its white fur collar, alighting from a snow-covered tractor accompanied by his frost-encrusted helper, and arriving at their door with a bulging sack.

The 1947 Christmas was one the children of the glen would never forget.

Metric Conversion Table

Wearne Engineering Pty. Ltd. of Thornleigh, N.S.W. manufactured chain diggers mounted on Howard 2000 tractors in the early 1970s. These, and the Wearne Hydrocranes, were designed by Ian Wearne and sold strongly to government utilities. The above metric conversion table was imprinted on an anodised plate as a promotional handout.

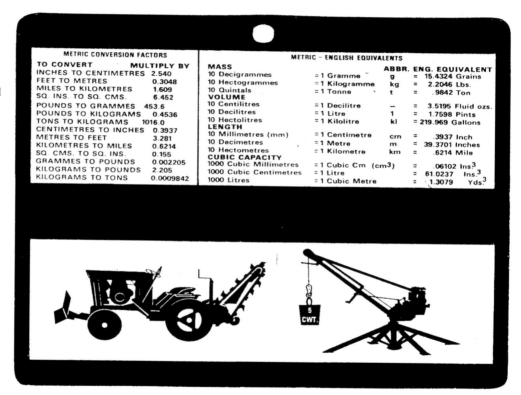

A 1936 International McCormick Deering T20 crawler, owned and restored by David Miller, Carisbrook, Vic. (Photo: D. Miller)

1963 Allis Chalmers ED 40. During the early 1960s, Allis Chalmers (U.K.) made a bulk purchase of unwanted standard Ricardo 4 cylinder diesel engines from Massey Ferguson. (The troublesome engine had been used in the M.F. 35 but subsequently replaced by the Perkins 152D 3 cylinder unit.) A new model Allis Chalmers ED40 was created, wrapped around the standard Ricardo engine. The stylish tractor automatically inherited the starting problems that had been experienced in the M.F. 35 with the same engine. The ED 40 was sold for a short period in Britain and Canada before it was quickly and quietly withdrawn from the market. The unit pictured is owned by Steve and Rachel Rosenboom of Pomeroy, Iowa. (Photo I.M.J.)

1937 Allis Chalmers WC. The steeled, wheeled version is equipped with unusual skeleton rear wheels. The WC was built between 1932 and 1948. It was one of the most successful of the Allis Chalmers range. The example pictured is yet another exhibit in the Rosenboom stable at Pomeroy, Iowa. (Photo I.M.J.)

Index of Illustrations

A.D.N., 255
Allis Chalmers WC, 266
Allis Chalmers ED 40, 264
Aultman Taylor 30 - 60, 118,119
Austin, 8
Aveling Porter 8 n.h.p., 7

Bailor, 87
Bean Row Crop, 97
Benz Sendling, 250

Caldwell Vale, 131
Case 10 - 20, 216
Case 500, 221, 222
Case LA, 220
Chamberlain Champion - Tail End Charlie, 45
Chamberlain Super 90, 142,143
Clark CA 1, 219
Clayton Chain Rail 35 h.p., 15, 16
Clayton Chain Rail 40 h.p., 17
Cletrac 15, 265
Clydesdale, 166
CO-OP 1, 11
CO-OP 2, 12
CO-OP 3, 13
CO-OP S3, 14
Common Sense 40 h.p., 227
Custom Gyrol Fluid Drive, 212

David Brown 2 D, 95, 96
David Brown 50D, 243
David Brown 50TD, 243
David Brown Cropmaster, 43
David Brown Cropmaster Diesel, 94
David Brown Super Cropmaster, 244
Deering Binder, 202
Dinkum Digger Mark 2, 141

Eicher EAKL 1509, 251
Eicher EM 200, 251
Emerson Brantingham Q, 207
Empire 88, 206
Federal Aircraft Hamilcar, 219

Fendt 6 h.p., 99
Fendt Farmer 2, 101
Fendt Geratetrager F 12 GT, 100

Fendt Holzgasschlepper G 25, 100
Fendt Xylon, 102
Ferguson assembly line, 228
Ferguson Model A, 144, 229–232
Ferguson TEA, 30
Ferguson TEA/ Disc Harrow, 27
Ferguson TEA/ Plough, 27
Ferguson TED, 29
Fitch D4, 76, 78
Fitch D4 Engine, 75
Fitch D4 Transmission, 77
Ford 9N, 146
Ford B, 18
Ford B Engine, 19
Ford B on Trailer, 19
Fordson E27N Major, 167, 246
Fordson E27N P6, 130
Fordson F, 10, 263
Fordson F - The Christmas Tractor, 259
Fordson F/Athens Plough, 22
Fordson N, 86, 124
Fowler 3 - 30, 193
Fowler 4 - 40, 194
Fowler Anti-Balanced Plough, 183
Fowler BB Ploughing Engine, 177
Fowler Gyrotiller, 184
Fowler Mole Drainer, 176
Fowler Motor Cable Ploughing Engine UA528, 178
Fowler Motor Cable Ploughing Engine UA529, 182
Fowler Steam Plough Set, 176
Fowler Turning Cultivator, 183
Friday, 48, 50

Galloway Farmobile, 149, 150, 151
Garner, 196
Glasgow, 165
Graham Bradley, 104
Graham Bradley Continental,, Engine, 105

H.S.C.S. R30-35, 9
Hanomag 1985 Industrial Machinery, 241
Hanomag Diesel 30 h.p., 237
Hanomag factory 1874, 237
Hanomag K 55E, 239
Hanomag R 28, 240
Hanomag R 40, 238
Hanomag Row Crop, 238

Hanomag Schneeschleuder, 240
Hanomag Z 25, 237
Hart Parr 16-30, 10
Hart Parr 18 - 36, 129
Holder A12, 252, 253
Holt 2 ton, 225
Horse Ploughing, 202
Howard Kelpie, 140, 246
Howard Platypus 30, 140
Huber 15 - 30, 211
Husking, 203

Illinois Super Drive, 223
International Farmall F12, 263
International Farmall M, 44
International Farmall M - Sheppard Diesel Powered, 41
International Farmall Super A, 245
International Friction Drive, 198, 199
International Mogul Type C, 30, 33
International Mogul Type C - remains, 34
International Mogul Type C - restored, 35
International T 20, 261
International W 30, 85
Ivel, 201

Jelbart No. 123, 133
John Deer Dain, 171
John Deere - original prototype, 170
John Deere 8200, 174
John Deere B, 6
John Deere D, 173
John Deere Lanz 440, 72, 88

KL Bulldog, 129, 138
Kubota L 210, 90
Kubota M 8580, 91
Kubota T 15, 90

L.H.B. LHS 25, 254
Landini 1935, 189
Landini 8860, 191
Landini L 55, 190
Lanz Bulldog Allrad, 62
Lanz Bulldog D 15 - 50, 67
Lanz Bulldog D 1706, 70
Lanz Bulldog DT, 70
Lanz Bulldog H, 71, 110
Lanz Bulldog HL 12, 59, 60
Lanz Bulldog HR 2, 63
Lanz Bulldog HR 5, 64
Lanz Bulldog HR 6, 64
Lanz Bulldog L, 69, 116
Lanz Bulldog N, 244
Lanz Bulldog P, 111
Lanz Bulldog Q, 72
Lanz Bulldog Saloon Tractor, 66
Lanz Bulldog Traffic, 65, 66
Lanz Factory 1870, 54
Lanz Felddank, 63
Lanz Futterschneidmaschine, 53
Lanz Landbaumotor - 1912, 56
Lanz Landbaumotor - 1917, 57, 58

Lanz Portable Steam Engine, 55
Lanz Steam Traction Engine, 55
Latil H 11, 47
Loyd Dragon, 2

Mail Coach 1909, 204
Marshall - Field Marshall Series 3, 159
Marshall - Track Marshall 90, 160
Marshall 544, 160
Marshall Colonial Class E, 156-158
Marshall Colonial Class F, 153
Marshall M, 159
Marshall MP6, 160
Massey Harris 102, 87
Massey Harris 1932, 84
Massey Harris 30, 246
Massey Harris 744D, 167
Massey Harris General Purpose, 214, 215
Massey Harris MH 1, 217
McDonald EB, 132
McDonald EB 12 - 20, 207
McDonald TWB, 126, 137, 137
McLaren 4 cyl. Diesel Engine, 186
McLaren Motor Cable Ploughing Engine, 185, 186
Munktells Type 25, 256

Nebraska Test Rig, 23

Oliver 1800, 213

Post Office Van, 258

Rally Scene - Nebraska, 242
Renault R3042, 245
Renault HI, 10
Ronaldson Tippett Super Drive, 137, 224

Same 190 Titan, 248
Same Single Cylinder, 248
Sheppard Diesel SD 2, 38
Sheppard Diesel SD 3, 38
Sheppard Diesel SD 4, 39
Sheppard Diesel SD 4 - rear view, 39
Silver King R 44, 208
Simpson, 243
Square Turn, 218
Stack Building, 203
Steyr Daimler 180, 249
Steyr Daimler 80, 249

Turner Yeoman of England, 125
Twin City 20 - 35, 123

Veb Trakto–renwerk Geratetrager RS 09, 97
Vickers Aussie, 128

Wallis 20 - 30, 127
Wallis Bear, 210
Wallis Cub Junior, 209
Waterloo Boy N, 172
Wendeler-Dohrn Motorised Plough, 237